Ecology and Man in Mexico's Central Volcanoes Area

Springer-Science+Business Media, B.V.

Ecology and Man in Mexico's Central Volcanoes Area

Edited by

Gerrit W. Heil
Department of Plant Ecology,
Utrecht University, The Netherlands

Roland Bobbink
Department of Landscape Ecology,
Utrecht University, The Netherlands

and

Nuri Trigo Boix
Departamento El Hombre y su Ambiente,
Universidad Autonoma Metropolitana, México

Springer-Science+Business Media, B.V.

A C.I.P. Catalogue record for this book is available from the Library of Congress

Additional material to this book can be downloaded from http://extras.springer.com.
ISBN 978-94-010-3756-3   ISBN 978-94-007-0969-0 (eBook)
DOI 10.1007/978-94-007-0969-0

Printed on acid-free paper

Contents

Preface

The main activities of the economically active population around The Iztaccíhuatl and Popocatépetl volcanoes region lie in the primary sector (65-90%). Of the people working in this sector, those dependent on agricultural or pastoral activities generally have an income significantly lower than the minimum wage in Mexico. Of the activities in the area, these agricultural, pastoral, and forestry activities probably have the most direct effect on the ecology of the volcanoes and its immediate surroundings. Traditional farmers, producing crops such as beans, pumpkins and cucumbers, generally work on small fields using traditional methods and animal traction. Modern farming, geared towards intensive production develops on larger sites making use of modern machinery, fertilizers, and pesticides. As the area under agriculture continues to increase every year, the attendant opening of large forested areas, soil modification, and ensuing erosion make it almost impossible for forest recovery. Extensive forestry in the region mainly concerns cutting and collecting wood, cutting *Pinus*-branches for torches or for utensils for open-fire cooking, collection of mushrooms, and hunting. Although these (often clandestine) activities seem to be small-scale, their adverse effects on the forest have been substantial. Weekend visitors from Mexico City heavily dominate recreation, with tourism concentrated near the roads leading to and inside the park. Lacking organization and facilities, most recreational activities have had significant environmental impact on the area

In many countries, the decline of nature has occurred because of changes in land use. Additionally, since the eighties of the last century, nature is increasingly threatened by environmental problems, such as by disturbance and pollution. For many years, changes in land use have obscured other changes in the environment. However, changes in nature are strongly delayed because of biotic and abiotic feedback mechanisms of ecosystems, such as regrowth.

The general public has recognized that environmental problems are likely to cause the current decline of nature. As a result of land use intensivation, but also because of fragmentation, less area have become available to plants species. However, the impact of management on natural ecosystems is not always clear. Following a set of guidelines in order to minimize adverse effects on nature is relatively simple, but to manage land use for multi-purposes is much more complex. Hence, the urgency develops to evaluate the effectiveness of management The adverse rapid changes in land use around the mountains make it essential to implement long-term management plans that must prioritise the conservation of the natural habitats. The urban area is extending both over the forest and over the agricultural lands and there is a displacement of the forested areas due to the opening of new lands for agriculture and grazing. Especially in the situation of the volcanoes, the agricultural lands have increased almost 50% in the last 15 years. This has not only caused the reduction of the forest areas but also its fragmentation. That is, the formation of vegetation "islands" which affects mostly the animal populations as a result of the reduction of their habitat. Reflecting this concern, in 1991 the Department of Man and His Environment of the Autonomous Metropolitan University at Xochimilco with the departments of Plant Ecology and Landscape Ecology, Utrecht University, began a collaboration project to develop a management plan for Iztaccíhuatl and Popocatépetl National Park and its surroundings.

The development of this plan included a series of three-week post-graduate courses during the period 1992 – 1999. During these courses part of the data were collected, the rest of the data was collected during numerous field campaigns by the authors, collaborators, and students. Together with the vegetation data and the data on different abiotic factors, the data on wildlife formed the backbone of the research on the natural system. The integration with the human system was based on a three-week course on the problem of scaling of socio-economic and ecological processes, i.e. what kind of knowledge is needed to analyse the problem in order to achieve the ultimate goal. For this, methodologies of Multi Criteria Decision Making - MCDM - were treated within the framework of Natural Resources Management.

"The neglect of ecology is the one most serious weakness of modern technology, and it goes hand-in-hand with our reluctance to be participating members of the whole community of living species."

Alan Watts, 1966. The Book - On the taboo against knowing who you are.

Gerrit W. Heil

Chapter 1

Mexico's central volcanoes area: an introduction

Roland Bobbink* and Gerrit W. Heil**
*Deparment of Landscape Ecology, **Department of Plant Ecology, Faculty of Biology, Utrecht University, The Netherlands

Key words: Volcanoes, diversity, vegetation, climate, soil

Abstract: Mexico, the third largest country of Latin America after Brazil and Argentina, is a long horn-like strip of land running northwest to southeast, sided by the Gulf of Mexico or Caribbean Sea in the east and the Pacific Ocean in the west. The highlands of southern Mexico are a geologically complex region separated by the Isthmus of Tehuantepec into two mountainous areas. Mexico has been placed within the top seven of most diverse countries of the world. The climatic variation across Mexico is high, and mostly related to the interplay of three influences, viz. the influence of the subtropical high-pressure area, the prevailing northeast trade wind, and the altitude ranging from 0 to over 5000 m. Most of the major biomes are found in Mexico. However, the biodiversity of Mexico is nowadays under serious threat because of the strong increase in population pressure. The central part of the Trans-Mexican Volcanic Belt rises to elevations clearly above 5000 m and forms the background of the Iztaccíhuatl-Popocatépetl National Park. Settlements spread over the whole region of the Valley of Mexico and Puebla, invading the foothills of the Iztaccíhuatl-Popocatépetl area increasingly. It is most likely that the increase in population pressure in the last decades of the 20th century have lead to changes in land use in the volcanoes area. These mountainous areas are essential water catchment areas, and are listed on the 1993 United Nations list of National Parks and Protected Areas. Although selected as a protected area, the park and its buffer zone are subject to significant disturbances, i.e. illegal tree felling, burning, and grazing, because of its proximity to Mexico-City and Puebla. At the end of this chapter a short introduction to the other chapters is given. In addition to the book, an accompanying CD with all Figures of the different Chapters in a series of PowerPoint files and a list of plant species found in the Iztaccíhuatl-Popocatépetl area is included.

1

G.W. Heil et al. (eds.), Ecology and Man in Mexico's Central Volcanoes Area, 1–18.
© 2003 Springer Science+Business Media Dordrecht

1.   MEXICO: AN OVERVIEW

Mexico, the third largest country (ca. 1.96 billion km^2) of Latin America after Brazil and Argentina, is a long horn-like strip of land running northwest to southeast (32^0 42' to 14^{0}30' N), sided by the Gulf of Mexico or Caribbean Sea in the east and the Pacific Ocean in the west (Fig. 1.1). About half of its territory is south of the Tropic of Cancer. Mexico is part of three crustal plates that helps to explain its diversity of landforms and its complex topography. A large part of Mexico involves a southern extension of the North American plate, ending at the Isthmus of Tehuantepec. It forms an elevated plateau of land, Mesa del Norte & Mesa Central, ranging from ca. 1,200 m in the north to around 2,300 m at its southernmost point in Mexico City. This large plateau is characterised by series of flat-floored basins intersected with low hills, often of volcanic origin. Some of the basins have rivers, which drain to the Gulf of Mexico or Pacific Ocean, however several of these basins drain internally with no outlet to the seas and contain shallow lakes or dry saltpans. The great Mexican plateau is framed by two major mountain chains running north-south, the Sierra Madre Occidental parallel to the Pacific coast and dropping steeply to a narrow coastal plain along the Pacific, and the Sierra Madre Oriental parallel to the coast of the Gulf of Mexico, forming a wider eastern coastal plain. A transverse chain of active and inactive volcanoes, the Trans-Mexican Volcanic Belt (*"Eje Neovolcanico"*), forms the southern border of the plateau. It crosses Mexico from west to east at the latitude of Mexico City (ca 19^{0}00' to 21^{0}00' N) and is characterised by many old cinder cones and several tall volcanic peaks. South of the Trans-Mexican Volcanic Belt, the Balsas river drains a structural depression.

The highlands of southern Mexico are a geologically complex region separated by the Isthmus of Tehuantepec into two mountainous areas, viz. the Sierra Madre de Sur in the south and the Chiapas highlands in the east, with tropical lowlands near both coasts. The Isthmus of Tehuantepec is the boundary between the North American and the Caribbean crustal plate. The end of the Mexican "horn" is formed in the east by the flat, limestone peninsula of Yucatán (< 500 m) (e.g. Ferrusquía-Villafranca 1993, Rees 1996).

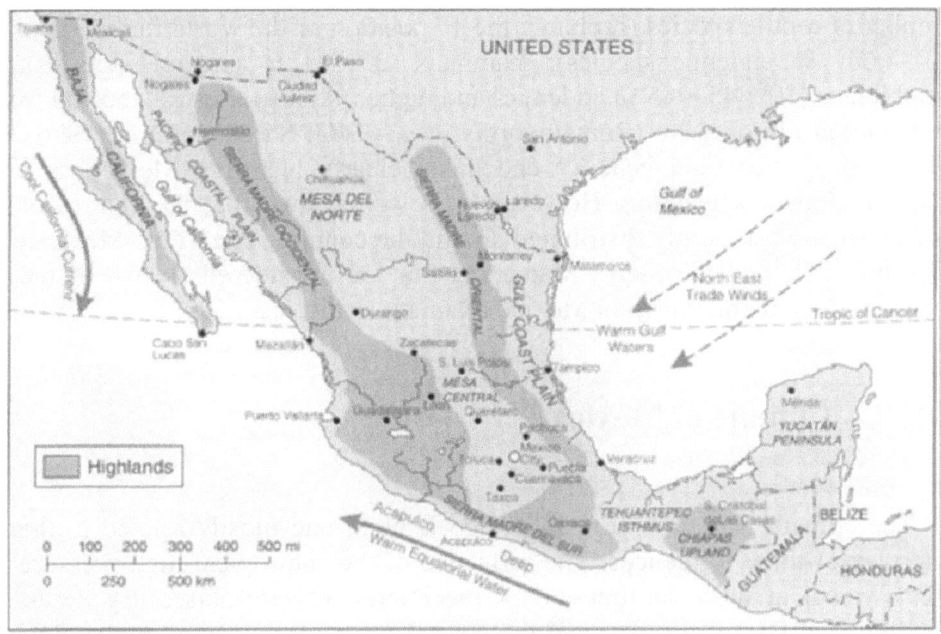

Figure 1.1. Overview of the geography of Mexico (redrawn with permission from Rees (1996)

2. MEXICO, A MEGA DIVERSITY COUNTRY

Mexico has been recognized as one of the most diverse countries in the world with roughly 10-12 % of world's species in a territory of only approximately 1.5% of the world's total emerged land. This pattern certainly holds for the better-known biological groups, e.g. plants, fungi, terrestrial vertebrates, and butterflies (e.g. Toledo and Ordóñez 1993; Sarukhán *et al.* 1996). Species endemism is also high, with percentages within groups ranging from 9 to 60 %. Because of this, Mexico has been placed within the top seven of most diverse countries of the world, after Brasil and Colombia, but ahead of Indonesia, Madagascar, Democratic Republic of Congo and Australia (Mittelmeier 1988). In general, the explanation for the high number of species richness in Mexico has been sought in the complex geologic history, varied climatic regions because of the large altitudinal differences and the fact that two of the major realms (the Nearctic and Neotropical) meet in Mexico (e.g. Sarukhán *et al.* 1996). Especially, the number of reptile species (probably the 1st position of the world ranking; ca. 50 – 60 % endemic species), mammals (2nd; 31 % endemic species), amphibians (10th; 45 – 55 % endemic) and higher plant species (4th; 55- 60 % endemic) are impressive (Ramamoorthy *et al.* 1993; Sarukhán *et al.* 1996). The high species (and generic) endemism clearly highlight Mexico as a major region of speciation. However, it is well known that the centres of endemism are not evenly distributed around the country. The Trans-Mexican Volcanic Belt and the Sierra Madre del Sur are considered as two of the major centres of endemism in Mexico (Ramamoorthy *et al.* 1993).

2.1   Climate of Mexico

The climatic variation across Mexico is high, and mostly related to the interplay of three influences. The influence of the subtropical high pressure area, typical at these latitudes, over the Pacific Ocean causes dry desert conditions in Baja California and the coastal lowlands surrounding the Gulf of California, but it affects two thirds of Mexico's land mass. A second influence is the prevailing northeast trade wind that comes from the Gulf of Mexico and brings most of the country's rainfall. Finally, altitude, ranging from 0 to over 5000 m, can strongly modify the otherwise (sub) tropical high temperature conditions, which dominate through lowland Mexico. In addition, it causes rain-shadow effects, which reduce precipitation on the Mexican plateau and the west-facing slopes, and leads to stratified ecological zones at the steep mountains slopes. Because of this all, four (A, B, C & E), out of five, of the main Köppen climate groups are found in Mexico (Mosiño Alemán and García 1974, Ramamoorthy *et al.* 1993).

The seasons in Mexico are defined by rainfall rather than temperature, unlike regions in temperate regions at higher latitudes (> 40°). The subtropical high-pressure area weakens on its southern side as the Intertropical Convergence Zone moves further north. Because of this, the northeast trade winds intensify to bring the rainy season from May to October. As the Atlantic subtropical high-pressure cell builds greater stability southwards, the trades weaken in winter and spring, bringing the dry season. The trade winds absorb humidity from the warm waters of the Gulf of Mexico. The flat Gulf coast has insufficient relief to cause topographic uplift and much rainfall. The high mountains of the Sierra Madre Oriental and Chiapas receive a high amount of rainfall, as the trades have to rise. Once over the mountain tops, the now drier air encompasses the Mexican plateau with much lower precipitation rates, from ca. 1040 mm in Mexico

City to 430 mm in Chihuahua in the North. As the air descends to the Pacific, it hardly releases moisture. The west sides of the mountains are thus drier and get their precipitation from Pacific storms, which frequency increases to the south (e.g. Mosiño Alemán and García 1974, Rees 1996).

2.2 Vegetation of Mexico

The richness of vegetation types in Mexico is overwhelming. Most of the major biomes are found in Mexico. Apart from Mexico, only India and Peru have a similar diversity of plant cover. The highly diverse vegetation of the semi-arid parts of Mexico can only paralled by the vegetation of South Africa (Rzedowski 1993). Many local or regional vegetation types have been described for Mexico, but six major continental ecological zones (with their typical vegetation types) are generally distinguished (Table 1.1). From xerophytic desert vegetation to tropical evergreen rain forests, but also mountainous cloud forests, temperate Conifer-Oak forests and even alpine grasslands communities can be found in the territory of Mexico. With respect to their area, cloud forest and tropical evergreen forest have the highest species richness of plants, but the arid / semi-arid (xerophytic scrub and grassland) and the sub humid temperate zone (conifer and oak forests) have especially high percentages of species endemism (> 50%), whereas only 5 % of the plant species of tropical evergreen forest is endemic.

The sub humid temperate zone (Cw climates) covers the greatest part of the mountains of Mexico (> ca. 2000 m) and it is (or was) characterised by conifer (especially *Pinus* spp.), oak and mixed forests. This more or less insular "ecological zone" is of very high importance for the biodiversity of Mexico, because of its species richness and, especially, very high proportion of endemic species in flowering plants, tree species (a.o. *Pinus* & *Quercus*), mammals, amphibians, and reptiles (e.g. Ramamoorthy *et al.* 1993, Sarukhán *et al.* 1996).

The original area of this zone was estimated as 377,000 sq km, of which in 1992 ca. 68 % could still be classified with remotely sensed images as this forest type (Castilleja 1995). This also is the case for the Trans-Mexican Volcanic Belt, with one of the highest concentrations of endemism known at non-island conditions. However, the biodiversity of this zone is nowadays under serious threat because of the strong increase in population pressure (see chapter 3).

Table 1.1. Estimated proportion of the main vegetation type in Mexico's phanerogamic flora. Last column indicates percentages of endemic species in the total number of species per vegetation type (after Rzedowski 1993; Toledo and Ordoñez 1993)

Major vegetation type (Rzedowski 1993)	Ecological zones (Toledo & Ordonez 1993)	Relative cover in Mexico (%)	Number of species	Percent of Flora	Percentage of endemic species
Xerophilous scrub and grassland	Arid / semi-arid (Bs – Bw)	50%	6000	20	60
Conifer and oak forest	Subhumid temperate (Cw)	21	7000	24	70
Cloud (montane) forest	Humid temperate (A(C)m, C(A)m)	1	3000	10	30
Tropical evergreen forest	Humid tropic (Am & Af)	11	5000	17	5
Tropical subdecidious, decidious, & thorn forests	Subhumid tropic (Aw)	17	6000	20	40
Alpine (bunch)grasslands (Zacatonal)	Alpine (E)	< 1	?	?	High
Aquatic and subaquatic vegetation	Aquatic (azonal)	-	1000	3	15
Ruderal vegetation	Others	-	2000	6	20

3.   THE TRANS-MEXICAN VOLCANIC BELT AND THE IZTACCÍHUATL-POPOCATÉPETL AREA

As described earlier, the Trans-Mexican Volcanic Belt is a mountain range of active and inactive volcanoes that forms the southern border of the great Mexican plateau. It crosses Mexico from west to east at the latitude of Mexico city (ca $19^{0}00'$ to $21^{0}00'$ N) and is about 930 km long and 120 km wide (ca. 9% of Mexico). The topography is characterised by many old cinder cones and several tall volcanic peaks. The mountains enclose many basins, rivers and volcanic plateaus. The altitudinal range in the belt is 1,000 – 5,700 m, but the dominant altitude band is between 1,500 and 2,500 m. Three of the highest peaks in North America are present in this zone, namely Citlatépetl (Pico de Orizaba; 5,699 m), Popocatépetl (5,452 m) and Iztaccíhuatl (5285 m). Physiographically, most of the region has been formed by volcanism; this high volcanic activity is probably related with the subduction of various small tectonic plates under the North-American plate (e.g. Ferrusquía-Villafranca 1993).

The highest part of this volcanic belt became the centre of the Aztec civilisation in the 13th century, because of its temperate climate and fertile volcanic soils. The Aztec built their capital city, Tenochtitlán, on the islands of Lake Texcoco. After the destruction of the city by the Spaniards, Cortés established Mexico City on the same location. In the second half of the 20th century, it has become one of the most populated metropolitan areas of the World (Fig. 1.2).

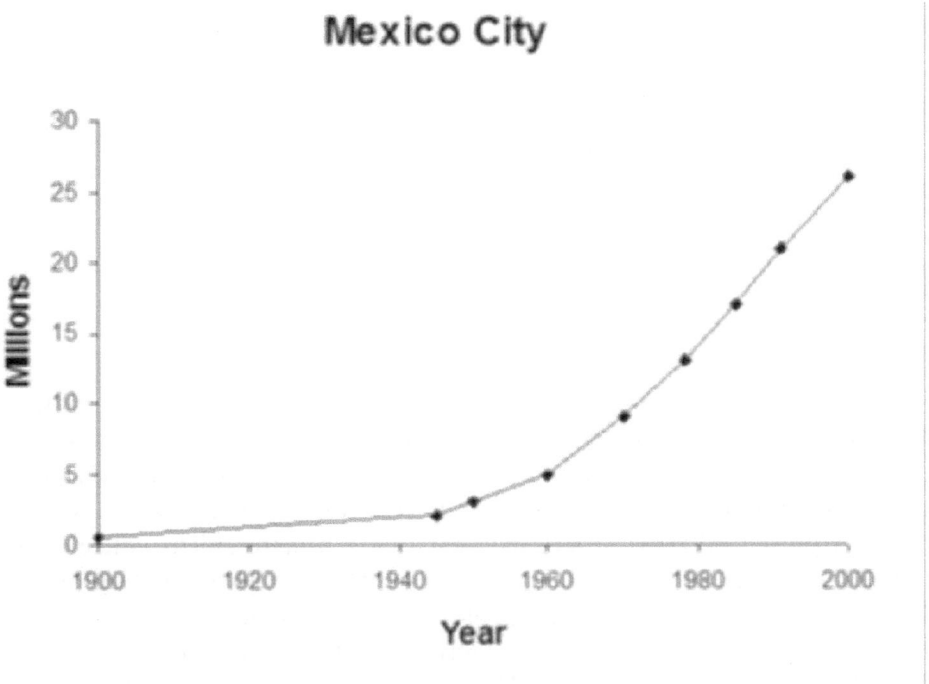

Figure 1.2. Population in the Mexico City metropolitan area (s.l.) in the 20th century (from http://www.sru.ed)

Fifty kilometres east of Mexico City and encompassing the states of Mexico, Puebla and Morelos, the snow-capped volcanoes of Iztaccíhuatl and Popocatépetl, the central part of the Trans-Mexican Volcanic Belt, rise to elevations clearly above 5,000 m and form the background of the Iztaccíhuatl-Popopatépetl National Park (commonly Iztac-Popo) (Fig. 1.3).

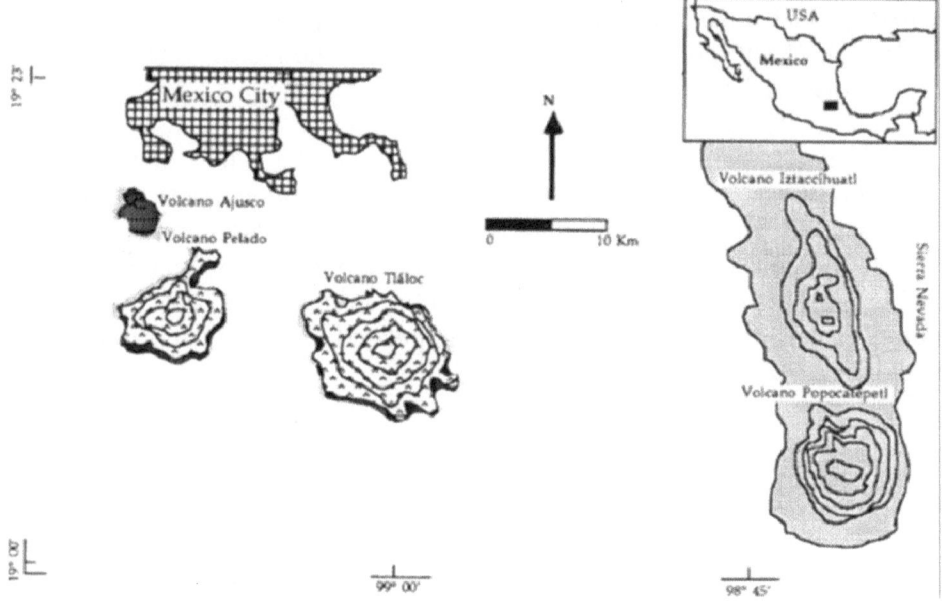

Figure 1.3. Overview of the Iztaccíhuatl-Popocatépetl National Park, its surroundings and its location in Central Mexico

3.1   Two volcanoes

Iztaccíhuatl ("sleeping woman" in Nahuatl; 5285 m) is an extinct strato-volcano formed of layers of viscous lava flows, flow breccias, and tephra. The common rock types are andesite and dacite. The volcano started to develop around 900,000 years ago and its growth was in two phases. In the older phase (900,000 – 600,000 years ago), a large shield volcano has built up with a summit caldera, with cones and lava flows erupting on the flanks of the shield. The younger phase (< 600,000 years) consisted mostly of lava flows and pyroclastic material erupted from the top of the Iztaccíhuatl and from its flanks. The volcanic activities ceased in the late Pleistocene, ca. 80,000 years ago (Velázquez-Selem 1997, Decker and

Decker 1998, http://volcan.und.edu). The present form of the volcano is highly eroded and hardly anymore related to the volcanic activities that caused its existence.

Popocatépetl (Nahuatl for "smoking mountain"; 5450 m) is a typical example of a strato volcano that stands ca. 3000 m above the surrounding plateau. Its snow-capped peak is 15 km south of the Iztaccíhuatl, and separated from it by the Paso de Cortés (3700 m). Popocatépetl has an almost perfect conical shape to an altitude of 5000 m, where the cone becomes more irregular, because of the Pico del Fraile, a remainder of the old strato-volcano that forms the base of the present Popo. Several heavy pyroclastic and lava eruptions of the Popocatépetl have occurred in the late Pleistocene and Holocene.

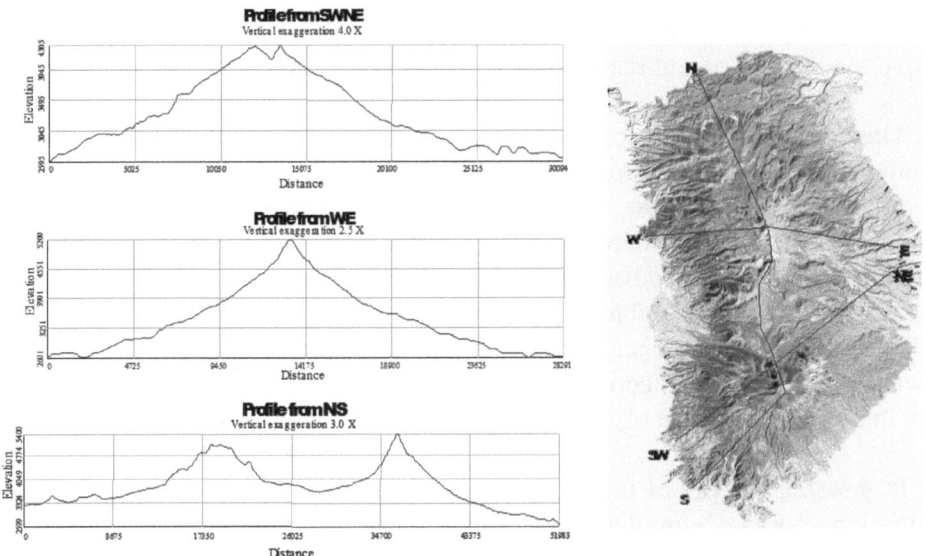

Figure 1.4. Altitudinal profiles through the Iztac-Popo area from SW – NE, from E – W, and from N – S (elaborated by Gilberto Hernández).

More recent, large plinian eruptions, like the Vesuvius eruption that destroyed Pompeii, in 3,100 BC, 400-800 BC and around 800 AD have taken place, which produced pyroclastic flows all around the volcano to distances of 10 km, thick ash deposits dispersed, especially in the direction of the valley Puebla and years of destructive mudflows (lahrs) filled this valley. Popocatépetl has had many small to moderate eruptions during the last 1000 years, especially with small explosions, local seismicity and ash fall. From December 1994 till present, several small to moderate eruptions with gas and ash emissions and some lava flows occurred (Palacios 1996; Decker and Decker 1998). An overview of the physiography of the Iztac-Popo Park, illustrating the form of both volcanoes, is given in Figure 1.4.

3.2   Soils

The soils in the Iztac-Popo National Park are thus generally formed on old or young volcanic materials, such as ashes, lahars and on lava flows. The most important soils in the area are: (i) andosols, (ii) regosols and (iii) litosols (Chávez and Trigo 1996). Andosols are the most common soil type in the Iztac-Popo area, especially in the eastern half. These soils are formed on relatively recent, base-rich volcanic deposits and characterised by a dark, fine, granular or crumby A layer and a well drained and blocky B layer, above the layered parent material of volcanic ashes (Fig. 1.5).

They are humid, light soils with a high content of a morph volcanic glass. Young andosols may have an AC profile (Bridges 1997). Regosols developed on top of loose volcanic material, without andic conditions and they are mostly very low in nutrients. Finally, lithosols are shallow (< 0.30 m) soils, which stay undeveloped because of the harsh and steep conditions at the great heights at which they occur (> 4500 m on the peaks of both volcanoes). Furthermore, cambisols are also present in the area, especially at lower altitudes and on steep slopes. They are relatively well-developed soils, low in nutrients because of weathering (e.g. Bridges 1997).

In general, the pH of the soils is relatively high (5 – 7), with high base saturation and an accumulation of organic material in the topsoil. At higher elevation extreme temperatures resulted in low microbial activity, causing very deep (to 1.5 m) organic layers, especially in the relatively flat parts of the Iztac-Popo area. Nutrient concentrations and C:N ratios are moderately high.

Figure 1.5. Typical volcanic soil on higher part (3500 m) of the Iztac-Popo area (photograph by R. Bobbink).

3.3 Hydrology and erosive effects

The water resources of the Iztac-Popo area are coming mainly from rainfall, which is quite high in summer, and melting snow or ice from the peaks of both mountains. This allows the formation of a few permanent streams and several temporal ones, especially in the rainy season. The hydrological systems of the area belong to two draining basins; the Basin of

Mexico and the Basin of the Balsas. The streams at the northwest side of the mountain ridge, north of 19°05' N, belong to the Basin of Mexico, eventually ending in the Gulf of Mexico. The streams south of 19°05' N, and on the east side of the chain drain to the Basin of the Balsas, and thus eventually to the Pacific (Chávez and Trigo 1996). The mountainous chain of the Iztac-Popo is in this way an essential water catchment area, especially for the valley of Mexico. Because of the large altitudinal differences and the relatively high precipitation in summer, long-term erosive processes considerably influence the landscape. Around the high peaks of the volcanoes, glaciers have formed moraines and caused the eroding of U-shaped valleys. Lower down, fluvial erosion has resulted in steep ridges and deep valleys or canyons (Van de Poll 1995).

3.4   Climate and vegetation

The local climate of the Iztac-Popo area is determined by its geographical location in Mexico (19 – 20° N) and by its altitude. The region has, as in many parts of Mexico, a rainy season in summer (May – October) and a dry season in winter (December – April) (Fig. 1.6). Mean annual precipitation range from ca. 900 to almost 1900 mm, and is clearly related with altitude (till 4000 m) and with the major orientation (east or west) of the volcanoes, because of the prevailing trade winds. The temperatures in the Iztac-Popo area are especially influenced by its elevation; the main part of the area is in the sub humid, mild to cool, temperate climate zone (Cw–two subtypes; 2500 – 4000 m). The lower part of this temperate zone has rather warm climate (e.g. Amecameca; Fig. 1.6), whereas above 3000 m much cooler conditions prevail (e.g. Hueyatlaco; Fig. 1.6).

A cold ("alpine") climate zone (E – according Köppen) is found above ca. 3900/4000 m, the altitude of the natural tree line at these latitudes (average temperature 3 – 5 °C; many frost days). Above ca. 5100 m, an extreme nival climate (F) with permanent snow or ice cover is present.

Across this large altitudinal gradient with its typical climate zones, a range of characteristic vegetation belts is found. The Iztac-Popo area is already lengthy known for its high alpine bunch grasslands ("zacatonal"), the characteristic coniferous (especially *Pinus* spp.) and mixed forests (Fig. 1.7) (e.g. Rzedowski 1978, Almeida *et al.* 1994, Chávez and Trigo 1996, Velázquez *et al.* 2000).

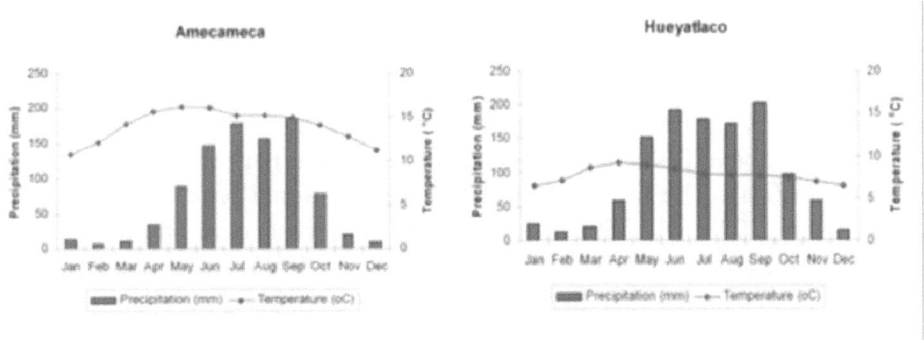

Figure 1.6. Diagrams of monthly precipitation and mean temperature in Amecameca (2470 m) and Hueyatlaco (3557 m; Paso de Cortes)

Figure 1.7. Photographic view of the slopes of the Volcanoes (forests - grasslands - summit)

Both the forests and grasslands are of major importance for the biodiversity of plant and animal life in Mexico, because of its central position in the Trans-Mexican Volcanic Belt where the Nearctic (boreal) and Neotropical realms meet and because of its large altitudinal range

(Rzedowski 1993; Toledo and Ordóñez 1993). E.g., it is one of the core areas in the restricted distribution of the endemic volcano rabbit (*Romerolagus diazi*) (Velázquez *et al.* 1996). Parts of the Iztac-Popo region became already a forest reserve in 1929, and the volcanoes and their surroundings above 3000 m were declared a national park in 1935, covering almost 60,000 ha. The limits of the Iztac-Popo national park were raised to 3,600 m in 1948 (37,350 ha) to facilitate forest exploitation in the northwestern parts of the area (Chávez and Trigo 1996). In general, it has, however, been emphasized that most of the higher parts of the area (> ± 2800) was in more or less natural condition at the beginning of the 20th century.

3.5   Threats

As in many developing countries, the urban population and its areas strongly increased in the last half of the 20th century. This certainly holds for the Valley of Mexico (see Fig 1.2) and the Valley of Puebla, nowadays with a population over 30 million or more together. The total surface of the urban settlements increased enormously in this period, while the area occupied with agriculture did not decrease, but even increased. Moreover, the settlements also spread over the whole region of the Valley of Mexico and Puebla, researching the foothills of the Iztac-Popo area more and more (e.g. López-Paniagua *et al.* 1996). It is thus likely that these changes in population pressure in the last decades of the 20th century have lead to changes in land use in the volcanoes area. Especially, an intensification of the use of the natural resources of the Iztac-Popo area can be expected. Most important in this respect is the shift from small-scale wood cutting for local use, to a more intensive slashing with modern machinery and a transition from traditional pastoral systems to a more intense system with higher frequencies of burning and higher grazing pressure, especially by cattle and sheep. In addition, recreational pressure also increased because of its proximity to Mexico City and Puebla (Chávez and Trigo 1996, Velázquez *et al.* 2000).

4.   AIMS AND SCOPE OF THIS BOOK

One could easily get the impression that the most effective way for man and nature to coexist is simply to assign them to different places. The last century has witnessed an explosion in the number of national parks, nature reserves, protected areas, or other tracts of the earth's surface similarly designated as exhibiting natural qualities worthy of some form of protection or preservation. Merely assigning people and nature to different places is not enough to protect nature. A boundary alone, a division of essentially zero width, is commonly insufficient to maintain the separation, to keep what is outside from penetrating the barrier and invading the protected. This holds almost without exception in the world. Reflecting this concern, in 1991 the Departamento El Hombre y su Ambiente (Department of Man and His Environment) of the Autonomous Metropolitan University at Xochimilco, in Mexico City, with World Bank financing, began a project (Chávez and Trigo 1996) to develop a management plan for the volcanoes areas around Mexico City.

Although natural areas can be distinguished from other areas according to a number of criteria (e.g., vegetation types), we will define and treat them here mostly in terms of land use and land cover. Ecology has demonstrated the importance of landscape structure—the spatial arrangement of different ecological communities and the nature of the connections among them in maintaining ecological integrity and sustainability (e.g. Forman 1990; Forman and Godron 1986). For example, the length of uninterrupted stretches of habitat is thought by many ecologists to be more important for many aspects of ecosystem preservation than the total area of such habitat.

In Chapter 2, we first describe the Iztaccíhuatl and Popocatépetl volcanoes region from a phytogeographical and a botanical point of view by their different vegetation communities occurring in the area as well as of their spatial distribution. Additionally, a description of most of the species present in the ground, herb and shrub layer of the main vegetation types as well as the tree-associated species has been made.

In Chapter 3, an analysis is given of the changes in natural communities in the Iztac-Popo area in the period 1986 to 1997 using satellite images and ground data of the vegetation. Especially, changes in land cover and the pattern of these changes are related to the observed changes in land use and the increase in disturbances in the area. The ecological consequences of these disturbances for the structure and function of the natural (forest) systems are delt with.

In Chapter 4, we make notice of the possibility that, through the interest for birds and bird watching, the mountain environments of the Iztaccíhuatl and Popocatépetl volcanoes can be conserved. We point out the most meaningful bird species for ecotourism as well as the places that can potentially offer to the visitor of the area greatest satisfactions within a bird watching activity. Ten routes along the northwestern part of the Iztaccíhuatl are recommended for bird watching.

In Chapter 5, a database including all mammalian species recorded in the region from 1839 up to 1997 is compiled. The records, at genus level, were linked to land cover classes obtained from TM Landsat satellite images taken in 1986 and 1997 in order to evaluate how land cover changes affect mammalian assemblages. It is shown that all natural land cover types provide habitat for different mammalian assemblages. The results show which cover types may have priorities for mammal conservation.

In Chapter 6, remotely sensed data in combination with field data are used to develop a spatially explicit model in order to answer the question of how spatially localised impacts such as collecting wood resources may have consequences for plant community assembly. The results show the interconnections and feedbacks between community assembly and stochastic disturbances.

In Chapter 7, insight is obtained into the effects of changes in land cover on the non-fossil carbon cycle and the time-scales over which they occur. For this a dynamic simulation model on carbon cycling of terrestrial ecosystems has been converted into a spatial (GIS) environment. The results show that remotely sensed information could be successfully used as a primary data source for the carbon model.

In Chapter 8, we describe the strategic planning efforts applied to the recreational area of Iztaccíhuatl-Popocatépetl National Park and one of its outlying municipalities. A basis to this is that conflicts arise in direct proportion to the number of stakeholders or interests involved in the decision-making process. Results obtained in this research met the demands posed by the problem. The resulting alternatives represent a feasible basis, suggesting productive activities that are both economically and ecologically viable.

Finally, a synthesis is given in Chapter 9. Additionally, the results are put together in a framework for Natural Resources Management.

This book shows numerous methodologies for processing data from natural areas, illustrating it in applications to Mexico's Iztac-Popo National Park. The book is written for graduate students, ecologists, and natural resource managers, and it therefore responds directly and primarily to their viewpoints and objectives. However, since ecosystem management and decisions about natural-area preservation take place within socioeconomic and political environments that are dynamic and plural, conflicts with those objectives are sure to arise. Therefore, Chapters eight and nine go beyond managers' objectives and viewpoints *per se* to explore and respond to potential conflicts.

In addition to the book, an accompanying CD with all Figures of the different Chapters in a series of PowerPoint files and a list of plant species found in the Iztaccíhuatl - Popocatépetl area is included.

REFERENCES

Almeida, L., Cleef, A. M., Herrera, A., Velázquez, A., and Luna, I. (1994). El zacatonal alpino del Volcán Popocatépetl, México, y su posición en las montañas tropicales de América. Phytocoenologia 22, 391-436.

Bridges, E. M. (1997). World Soils, 3rd edn. Cambridge University Press, Cambridge.

Castilleja, G. (1995). Mexico. In: The conservation atlas of tropical forest: the Americas, (C. S. Harcourt and J. A. Sayer, eds.), pp. 193-205. Simon and Schuster, New York.

Chávez Cortés, J. M. and Trigo Boix, N. (1996). Programa de manejo para el parque nacional Iztaccíhuatl-Popocatépetl. Universidad Autónoma Metropolitana, Unidad Xochimilco, México D.F.

Decker, R. and Decker, B. (1998). Volcanoes, 3rd edn. W.H. Freeman and Company, New York.

Ferrusquía-Villafranca, I. (1993). Geology of Mexico: A Synopsis. In: Biological diversity of Mexico; origins and distribution In: T. P. Ramamoorthy, R. Bye, A. Lot and J. Fa, eds., pp. 3-107. Oxford University Press, New York.

Forman R.T.T. and Godron, M. (1986). Landscape Ecology, New York: J.Wiley and Sons.

Forman R.T.T. (1990). Ecologically sustainable landscapes: The role of spatial configuration. In: I.S. Zonneveld and R.T.T. Forman (eds.), Changing Landscapes: An Ecological Perspective. Springer-Verlag, New York. pp. 261-278.

López-Paniagua, J., Romero, F.J. and Velázques, A. (1996). Las actividades humanas y su impacto en el habitat del conejo zacatuche. In: Ecología y conservación del zatachuche (A. Velázquez, F.J. Romero and J. López-Paniagua, eds.) pp. 119-132. Publicaciones científicas, Fondo de Cultura Económica, México DF.

Mittelmeier, R.A., and Mittermeier, G.C. (1998). Megadiversidad, p 498. CEMEX, Mexico City, Mexico.

Mosiño Alemán, P. A. and García, E. (1974). The climate of Mexico. In: Climates of North America (R. A. Bryson and F. K. Hare, eds.) pp. 345-405. Elsevier, Amsterdam.

Palacios, D. (1996). Recent geomorphologic evolution of a glaciovolcanic active stratovulcano: Popocatépetl (Mexico). Geomorphology 16, 319-335.

Ramamoorthy, T. P., Bye, R., Lot, A., and Fa, J. (1993). Biological diversity of Mexico; origins and distribution. Oxford University Press, New York.

Rees, P. 1996. Mexico. In: Latin America and the Caribbean: a systematic and regional survey (B. W. Blouet and O. M. Blouet, eds.) pp. 207-236. Wiley and Sons, New York.

Rzedowski, J. (1978). Vegetación de México. Limusa. México D.F. , 432 pp.

Rzedowski, J. (1993). Diversity and origins of the Phanerogamic Flora of Mexico. In: Biological diversity of Mexico; origins and distribution (T. P. Ramamoorthy, R. Bye, A. Lot and J. Fa, eds.), pp. 129-144. Oxford University Press, New York.

Sarukhán, J., Soberon, J., and Larson-Guera, L. (1996). Biological conservation in a high beta-diversity country. In: Biodiversity, science and development: towards a new partnership (F. Di Castri and T. Younès, eds.) pp. 246-248. CAB International.

Toledo, V. M., and Ordóñez, M. J. (1993). The biodiversity Scenario of Mexico: a review of terrestrial habitats. In Biological diversity of Mexico; origins and distribution (T. P. Ramamoorthy, R. Bye, A. Lot and J. Fa, eds.), pp. 758-777. Oxford University Press, New York.

Van de Poll, H.M. 1995. Remote sensing based vegetation mapping of the National Park Iztaccíhuatl-Popocatépetl, Mexico.Faculty of Biology, Universiteit Utrecht, report nr 951102.

Vazques-Selem, L. (1997). Late quaternary glaciations of Teyotl volcano, central Mexico. Quarternary International 43/44, 67-73.

Velázquez, A., Romero, F.J., and López-Paniagua, J. (1996) Ecología y conservación del zacatuche. Publicaciones cientificas, Fondo de Cultura Económica, Mexico, D.F.

Velázquez, A., Toledo, V. M., and Luna, I. (2000). Mexican temperate vegetation. In: North American terrestrial vegetation (M. G. Barbour and W. D. Billings, eds.), pp. 573-592. Cambridge University Press, Cambridge.

Chapter 2

Classification and mapping of the vegetation using field observations and remote sensing

Nuri Trigo Boix*, Aurora Chimal Hernandez*, Gerrit W. Heil**, Roland Bobbink**, and Betty Verduyn**

*Departamento El Hombre y su Ambiente, Universidad Autonoma Metropolitana, México. **
Faculty of Biology, Utrecht University, The Netherlands*

Key words: classification, remote sensing, plant communities

Abstract: The Iztaccíhuatl and Popocatépetl volcanoes region is very important both from a phytogeographical and a botanical point of view due to their location within the Neovolcanic Transversal Axis. The combination of different factors favours the development of different vegetation communities that represent an important floristic richness. To be able to design and implement adequate management and conservation plans for the Iztaccíhuatl and Popocatépetl volcanoes area, it is necessary to obtain more knowledge of the different vegetation communities occurring in the area as well as of their spatial distribution. The method applied to accomplish a vegetation classification for the area considered in the first place the gathering of field data on the plant species composition of the various forests, and secondly a TM satellite image that has been used to process a supervised classification for the area. The potential of remotely sensed satellite imagery is limited to achieve a very comprehensive supervised classification. However, vegetation classes that cover a large homogenous areas (> 1 ha) are separated through a spectral analysis, which resulted in a map that distinguishes seven major vegetation classes. Additionally, a description of most of the species present in the ground, herb and shrub layer of the main vegetation types as well as the tree-associated species has been made. An updated species list is available on the c.d.

G.W. Heil et al. (eds.), Ecology and Man in Mexico's Central Volcanoes Area, 19–48.

1.   INTRODUCTION

The Iztaccíhuatl and Popocatépetl volcanoes region is very important both from a phytogeographical and a botanical point of view. This is due to their location within the Neovolcanic Transversal Axis, which includes the highest mountains in México, and to a marked altitudinal gradient. The combination of these factors favours the development of different vegetation communities that represent an important floristic richness. Previous research in the region has been either too extensive and coarse (Ern 1972; Fuentes 1975); or it has focused on very specific communities (Guzmán 1966, 1972; Madrigal 1967, Obieta and Sarukhan 1981; Cárdenas 1987) or locations (May-Nah 1971; Anaya *et al.* 1980).

Rzedowski (1975), when comparing the different vegetation communities of Mexico, mentions that the conifer and oak forests (highly represented in the volcanoes) are the plant communities with the higher number of species. Another community present in the area that is very important for depicting several endemic species, is the alpine vegetation (Almeida-Leñero *et al.* 1994; Escamilla 1996). In contrast, if we consider the area comprised between the 2,500 to the 2,800 meters above sea level (m), we have another rare and important vegetation community that is very similar to the mountain cloud forest *sensu* Rzedowski (1970). This community is present only in discontinuous gaps and under very special environmental conditions. In all of these communities we can find both Holarctic as Neotropical species and as a result of this diversity of communities in the area we can find approximately 45% of the 2,071 species which Rzedowski and Rzedowski (1985) recognized for the Valley of Mexico (Chávez and Trigo 1996).

From an ecological point of view, the area represents an important catchment for water retention, both for the Valley of Mexico as for the western area of Puebla. This includes the High Balsas Watershed, because the vegetation that covers the volcanoes increases significantly the water retention capacity. During the past decades, the vegetation communities of the volcanoes have survived in an acceptable conservation condition; which is fortunate when we consider that these communities are the habitats for a big number of animal species.

The adverse rapid changes in land use around the mountains make it urgent to implement long-term management plans that must prioritize the conservation of the natural vegetation habitats. The urban area is extending both over the forest and over the agricultural lands and there is a displacement of the forested areas due to the opening of new lands for agriculture and grazing. Especially in the situation of the volcanoes, the agricultural lands have increased almost 50% in the last 15 years (Chávez and Trigo 1996). This has not only caused the reduction of the forest areas but also it's fragmentation. That is, the formation of vegetation "islands" which affects mostly the animal populations as a result of the reduction of their habitat (see Chapter3). Therefore, adequate management and conservation plans for the forests are needed to establish buffer zones to limit this habitat fragmentation effect.

To be able to design and implement adequate management and conservation plans for the Iztaccíhuatl and Popocatépetl volcanoes area, it is necessary to obtain more knowledge of the different vegetation communities occurring there as well as of their spatial distribution. For this purpose, this study has been carried out to accomplish a vegetation classification of the Iztaccíhuatl and Popocatépetl volcanoes, which integrates the underlying processes that determine the distribution of the various vegetation communities at different scales.

2. METHODS

The method applied to accomplish a vegetation classification for the area is described in Figure 2.1. It considered in the first place the gathering of field data on the plant species composition of the various forests. From August 1994 to July 1997, vegetation data was gathered at 223 different sampling points in the volcano area. For all sampling points the position in UTM was obtained and data gathered about the dominant species, in most cases (189) including the percentage of canopy cover and in the rest (34) only the name of the dominant species.

In line with this information, the field data has been used for the classification of the vegetation according to the following types:
Abies religiosa forests, *Pinus hartwegii* forests, *Pinus montezumae* forests, *Cupressus* spp. forests, mixed conifer forests, broad-leaved forests, mountain cloud forests, mixed forests with conifer dominant species, mixed forests with

broad-leaved dominant species, alpine grassland, shrubs or secondary vegetation, induced grasslands, and crops.

Besides the former, the following non-vegetated areas can be found in the area, i.e. bare soils, urban areas, bare volcano peaks, and snow on the top of the volcanoes.

Figure 2.1. Overview of the data processing

Another essential information for this classification was a Landsat TM satellite image from March 1997 that was processed for radiometric correction. The geometric correction was done with the use of a 1:25 000 topographic map; eight reliable points could be found on both the images and the map for this purpose.

A preliminary classification (unsupervised) of the vegetation was used from the image for the field survey points. This was done for Bands 1 through 5, and Band 7. An exploratory data analysis was undertaken in order to determine which of the vegetation communities listed above could be separated with enough confidence to create a signature file with a training data set.

Finally, a supervised classification was obtained for the area using the maximum likelihood hard classifier MAXLIKE from IDRISI (Eastman 1993). This process was done in a recursive way by adjusting the signature file with respect to the information obtained from the exploratory data analysis and from field data gathered in a validation field trip carried out in October 1998.

3.   RESULTS

Based on our field knowledge of the area and the several years of field sampling, the vegetation and land use types of the Iztaccíhuatl and Popocatépetl volcanoes has been pre-classified as follows:

Natural vegetation
One-species conifer forest
     Abies religiosa
     Pinus hartwegii
     Pinus montezumae
     Pinus leiophylla
     Pinus ayacahuite
     Pinus teocote
     Pinus rudis
     Pinus pseudostrobus
     Cupressus spp.
Mixed conifer forests
Broad-leaved species forests
     Quercus spp.

Alnus spp.
Mountain cloud forest
Mixed forest
>   Conifer dominant forest
>   Broad-leaved dominant forest

Grasslands
>   Alpine
>   Sub-alpine

Degraded areas
>   Secondary vegetation
>   Induced grasslands
>   Bare soil

Man made landscapes
>   Crop
>   Urban

Volcano peaks
>   Bare rock
>   Snow

In spite of this information on the vegetation, the potential of remotely sensed satellite imagery is limited to achieve such a comprehensive supervised classification. Therefore, the results here presented are divided in two sections: the first one describing the supervised classification and the second one describing the vegetation composition of major vegetation types.

3.1   Supervised classification

Among the vegetation classes, only those that cover somewhat large homogenous areas (> 0.1 ha) can be separated through a spectral analysis from a remotely sensed image, and even for those, the spectral separation is not enough to distinguish among some of the vegetation classes. Such is the case, for example of the signature values obtained for *P. hartwegii* which are the same we can have in other types of conifer forests where densities are relatively low (less than 40%). Also, we find this closeness in signature values between mixed conifer/broad-leafed forests and forests where we find only conifers but with a rather dense cover (over 60%).

On the other hand, *Abies religiosa* forests are a special case where the signature (especially in band 4) can be distinguished from the rest. If we add to this the fact that this kind of forest is clearly distributed along ravines, covering relatively wide areas, then we can separate this class from the rest through the supervised classification procedures described in the previous section. As a result of this, some classes which share the same signature values are put together according to the vegetation classes depicted in Figure 2.2.

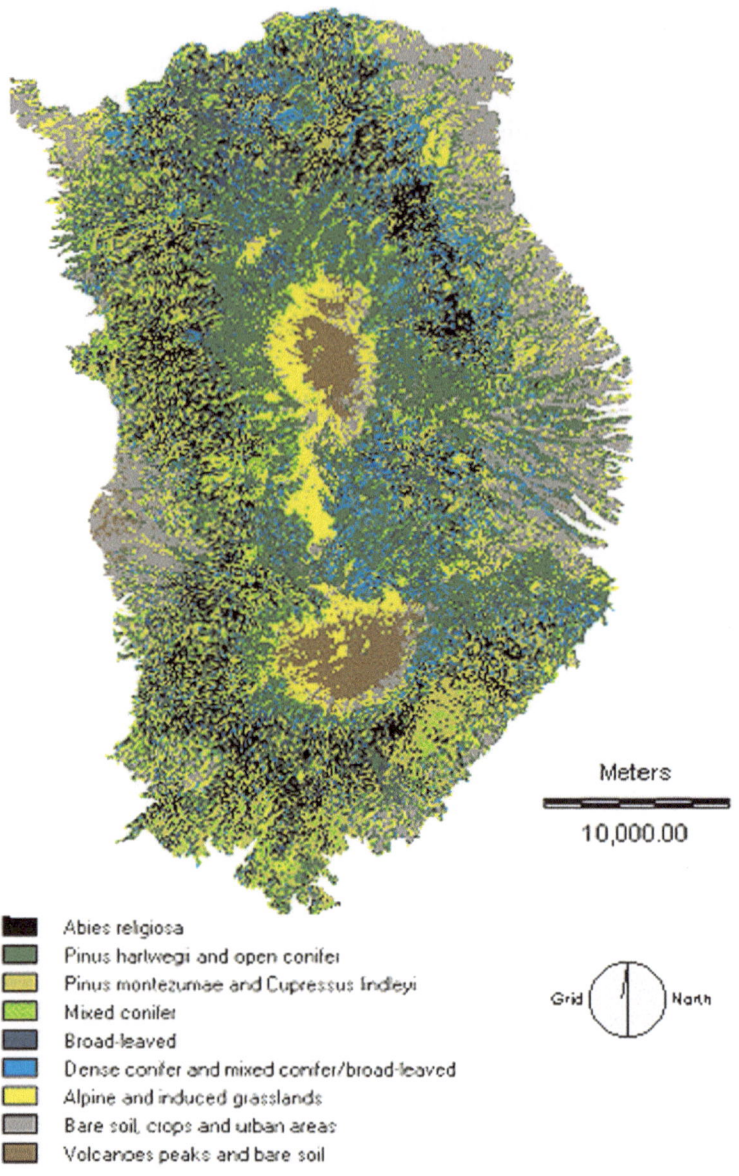

Abies religiosa
Pinus hartwegi and open conifer
Pinus montezumae and Cupressus lindleyi
Mixed conifer
Broad-leaved
Dense conifer and mixed conifer/broad-leaved
Alpine and induced grasslands
Bare soil, crops and urban areas
Volcanoes peaks and bare soil

Meters
10,000.00

Grid   North

Figure 2.2. Vegetation map (1997) of the Iztaccíhuatl - Popocatépetl area

3.2   Vegetation description

This section describes most of the species present in the ground, herb and shrub layer of the main vegetation types as well as the tree-associated species. A species list updated from the one published in Chávez and Trigo (1996) is added at the end of this Chapter. Families, genera and species are listed in alphabetical order following the classification from Mabberley (1998).

3.2.1   Fir forests

This is a perennial forest of the conspicuous species *Abies religiosa*. The tree cover is usually dense with relatively tall trees (Figure 2.3). As it reaches its altitudinal limit, the forest becomes more open, which also happens at the lower limits where the ravines begin. The altitudinal range for this community is from 2,700 to 3,550 m within deep andosol soils which are humid and rich in organic material due to the density of the forest and because of their location along the ravines. The weather corresponds to a temperate subhumid with summer rainfall (1,000 a 1,400 mm) and a mean annual temperature ranging from 7.5° to 13.5° C.

The structure of the vegetation, depending on the degree of disturbance, consists of two to three layers. The shrub and herb layers are rather poor in their number of species but they can also be very dense with respect to some of the Compositae, Leguminosae and Solanaceae species. According to the degree of humidity, the ground layer can have several moss species.

Typical samples of this forest can be found in the Cañada de Chopanac and in the Cañada of Tzotquinzinco as well as in Nexcoalanco. Mixed with this kind of community at lower parts we have *Cupressus lindleyi* and *C. lusitanica* both of great importance in the area distributed in a very patchy way near many of the fir and conifer forest areas.

In the lower tree layer, the most common accompanying species, which might have altogether a dominance of up to 80%, are *Clethra mexicana*, and *Arbutus xalapensis*. With increasing altitude, other less dominant accompanying species found at the high tree layer are *Pinus montezumae* in altitudes higher than 2,900 m, as well as *Garrya laurifolia* and *Quercus laurina*. The canopy cover of these tree species is usually close to 10%, when they are present and it becomes higher as the fir forest gives place to the pine and mixed types of vegetation.

Figure 2.3. Typical Fir forest with herb layer (photograph by N. Trigo)

In the shrub layer, which might have a cover of up to 70%, we can most commonly find the following species:

Fushia microphylla, Roldana angulifolia, Roldana barba-johannis, Senecio angulifolius, Senecio tolucencis, Lupinus montanus, Salvia elegans, Salvia gesneriflora, Physalis stapeloides, Cestrum thyrsoideum, Symphoricarpus microphyllus, Ceanothus coeruleus.

Two important endemic species of this kind of forest found between 2,600 and 3,200 m are *Ageratina parayanum* and *Ageratina isolepis*.

Under more open canopies (tree canopy cover ranging from 15 to 40%), *Verbesina virgata* is very common. In contrast, under closer canopies (tree canopy cover over 60 %), it is common to see *Smilax moranensis*, a "climbing" species.

In the herb layer, we usually find *Cirsium ehrenbergii, Trisetum virletii, Erigeron galeottii, Stachys nepentifolia, Packera sanguisorbae, Bromus anomalus* and *Deschampsia*. In open areas, characteristic species are *Geranium potentillaefolium, Castilleja moranensis, Erigeron galeotti, Echeandia reflexa, Oenothera roseum, Festuca amplissima, F. orizabensis* and *F. tolucensis*. In more disturbed areas, it is common to find *Ceanothus coeruleus, Castilleja moranensis* and *C. tenuifolia, Erigeron galeotti, Echeandia flavecens, Oenothera roseum, Bacharis conferta, Penstemon gentianoides, Brachypodium mexicanum* and *Vulpia myurus*; this latter species is abundant in the edge of roads.

In the ground layer with a cover of up to 40%, *Alchemilla procumbens* is very common together with *Eryngium carlinae*. On the other hand, in very humid places, we can find *Sibthorpia repens, Rubus liebmannii, Cinna poeformis, Arenaria bourgaei, Stevia* spp., *Agrostis* spp., *Eryngium* spp. and various moss species.

In the highest parts of this forests (3,500 m), *Abies religiosa* is associated with *Pinus hartwegii* in the upper tree layer and with *Salix oxylepsis* in the lower tree layer. The shrub layer is dominated by the following species: *Roldana angulifolius, R. barba-johannis, Rubus* spp. and *Ageratina* spp. The most common herb species present at this altitude associated with the fir forests are: *Muhlenbergia macroura, Penstemon gentianoides, P. roseus, Castilleja moranensis, Stevia* spp, *Festuca tolucencis, Lupinus montanus, Trisetum virletti, Bromus* spp, *Geranium*

potentillaefolium, Stipa ichu, Acaena elongata, Brachipodium mexicana,
and *Festuca amplisima.*

The ground layer is quite similar to that of open areas induced
grassland, with dominance of *Alchemilla procumbens, Potentilla
candicans, Poa annua, Eryngium carlinae, Acaena elongata, Deschampsia
pringlei, Muhlenbergia repens, Cyperus sesleroides* and *Carex leucophylla.*

3.2.2   Pine forests

These perennial forests are located between the 2,850 and 4,000 m with
a mean annual temperature ranging from 8 to 16 ° C and a mean annual
rainfall of 700 to 1,200 mm. These forests have the capability of
establishing in various types of soil including those that are poor in
nutrients or have a low depth. Some communities develop rather well on
rocky soils, as is the case of the *Pinus hartwegii* forest. These communities
are also found in vast areas of deep soils.

The most important and abundant of the pure pine forests is the one
formed by *Pinus hartwegii* which is worthy of a separate description since
it contains a characteristic set of associated species that are almost unique
to this type of forest.

Other most important pure pine forests is the one formed by *Pinus
montezumae* which shares many of it's shrub and herb species with the rest
of the pine forests and is found in rather big patches around both volcanoes.
Of less extension but also important are the forests of pure *Pinus
leiophylla, Pinus ayacahuite, Pinus pseudostrobus, Pinus teocote* and *Pinus
rudis.* The size of the patches of pure *P. hartwegii* and *P. montezumae*
should allow them to be recognised by a separate spectral signature in the
classification from the remote Landsat TM image but as will be discussed
in the next section, this is not so straightforward. The rest of the pine
species that are distributed in relatively small patches are engrossed within
the category of conifer forests.

In general, pine forests present trees with a height ranging from 20 to 30
meters. There is usually also a less high tree layer ranging from 10 to 20

meters formed by broad leaf species such as *Buddleia parviflora, B. cordata, B. sessiliflora, Alnus jorullensis, Arbutus* spp. and *Salix* spp.

The shrub layer of these forests is dominated by the following species: *Lupinus montanus, Fuschia microphylla, Ribes ciliatum, Ageratina* spp, *Salix paradoxa, Eringium* spp. and *Senecio* spp. A secondary shrub also found in pine forests is *Coriaria ruscifolia*.

In the herb layer we can find:
Erigeron galeoti, Festuca amplissima, F. tolucensis, F. orizabensis, Muhlenbergia macroura, M. quadridentata, , Calamagrostis tolucensis, Alchemilla procumbens, Arenaria spp., *Stipa ichu, Lupinus* spp., *Penstemon* spp., *Bidens bigelovii, Gamochaeta americana, Gnaphalium oxyphyllum, Stevia viscida, S. ovata, Galium ascherbornii, Salvia lavanduloides*.

Herbs which are characteristic of the *Pinus montezumae* forests are *Stenanthium frigidum, Gnaphalium liebmanii, Acaena elongata, Muhlenbergia rigida, Cinna poeformis, Castilleja arven*sis and *Alchemilla procumbens*.

A special reference ought to have *Festuca rzendowskiana*, an endemic species found between 2,750 and 3,450 m and reported by Rzedowski and Rzedowski (1979).

These pine forests also have a number of trailing species among which we can mention as exceptional *Sibthorpia repens* and *Didymaea alsinoides*.

Pinus hartwegii forests

Being the most distinctive of the pine forests, *Pinus hartwegii* is located between the 2,900 and 4,000 m and marks the upper limit of the forest vegetation in the mountains. This species is located in a wide variety of landforms including very steep hillsides with slopes of over 45 degrees. The type of soil can either be very deep or rocky although in the latter, the height of the trees is lower than 5 meters whereas in better soil conditions they can grow as high as 24 meters.

In general these forests are rather open with canopy densities ranging from 10 to 50 %. The shrub layer is not always present and when present, *Salix oxilepis, Senecio procumbens, Senecio mairetianus* and *Vaccinium geminiflorum* commonly dominate it. On the grassland edges towards the lower parts of these communities, it is very common to find *Juniperus monticola* almost forming a community by itself associated with grasses.

The herb layer, on the other hand, can reach canopy cover of up to a 40%. The most commonly associated species in the *Pinus hartwegii* forests are the following grasses (Figure 2.4):

Calamagrostis tolucencis, Muhlenbergia macroura, Festuca tolucencis, F. orizabensis, F. livida, Poa orizabensis, Agrostis vinosa.

Figure 2.4. *Pinus hartwegii* with the tussockgrass *Muhlenbergia macroura* in front
(photograph by A. Chimal)

Other types of herb species commonly found are *Alchemilla procumbens, A. pinnata, A. vulcanica, Eryngium proteiflorum, Acaena elongata, Penstemon gentianoides, P. roseus, Selloa plantaginea, Hieracium mexicanum, Lupinus montanus, Ageratina* spp., *Stevia* spp.,

Viola spp. and *Sedum* spp. In the more humid areas of these forests, it is common to find *Cinna poeformis* and *Stenantiun frigidum.*

An important species associated with this kind of forests and which Rzedowski and Rzedowski (1979) have considered as endemic of the valley of Mexico, is *Agrostis calderoniae*, found on the western slope of the Iztaccíhuatl.

A particularly dense area of *Pinus hartwegii* forest can be found in the higher parts of the San Juan Tetla Station on the Northeast part of the Iztaccíhuatl within the altitudinal range of 3,390 to 3,400 m. The shrub and herb layers are particularly rich for this kind of forest and we can also find species such as:
Senecio cinerarioides, Senecio sinuatus, Bacharis conferta, Phacelia platycarpa, Festuca amplissima, Ageratina glabratum, Erigeron carlinae, Geranium potentillaefolium, Brachiaria sp., *Rumex* sp., *Trifolium* sp., *Cyperus* sp., and *Bidens* sp.

3.2.3   Cypressus forests

Although *Cupressus* spp. is strongly associated with *Abies religosa*, in some areas it can be found in pure patches; a most important example of this is found on the southeastern slope of the Popocatépetl. Other important tree species sometimes present with very low densities in this kind of forest are *Arbutus xalapensis, Quercus rugosa* and *Q. laurina.*

Important shrubs in these forests are the following:
Verbesina oncophora, Symphoricarpus macrophyllus, Fuschia microphylla, Clematis diodica, Piqueria trinervia, Monina cililata, Baccharis conferta, Coriaria riscifolia, Salvia prunelloides, Senecio sinuatus, Valeriana clematitis.

Within the herbs group the following species occur: *Satureja macrosperma, Acaena elongata, Bidens anthemoides, Gamochaeta americana, Festuca rosei, Physalis chenopodiifolia* and *Stevia monardifolia.*

Because of the high degree of humidity found in this kind of vegetation, we find some creeping species such as *Smilax moranensis*, *Bromus anomalus* and *Galium aschenbornii*; ferns such as *Adiantum andicola*, *Asplenium monanthes* and *Driopteris wallichiana* Subsp. *Wallichiana*. At ground level, we find the trailing species *Didymaea alsinoides*.

3.2.4   Oak and alder forests

These broad-leaved forests are located between the 2,350 and 2,800 m. They are found both in deep as in shallow soils. The annual average temperature ranges from 9° to 13° C and annual rainfall from 700 to 1,200 mm.

The height of the tree layer and the phenology vary according to the oak-associated species (Figure 2.5). In general these communities are not very tall and depending on the depth of the soil and the slope they can reach up to 20 meters high.

The composition of the species includes some areas of predominantly deciduous species and others completely evergreen. From the 8 oak species reported in the area *Quercus crassipes*, *Quercus laurina* and *Quercus rugosa* are the more common. *Q. crassipes*, is usually found in areas combined with low-density communities of *P. montezumae* or with *Cupressus* spp.; associated species in the lower tree layer are *Garrya laurifolia*, *Arbutus xalapensis* and *Clethra*. In the shrub layer we commonly find the following: *Stevia*, *Dhalia*, *Baccharis*, *Ageratina*, *Laumouroxia* and *Valeriana*. Among these, an important endemic species is *Ageratina isolepis*. The herb layer is rich in individuals from the following genera: *Thalictrum*, *Galium*, *Geranium*, *Muhlenbergia*, *Penstemon* and *Salvia*.

Figure 2.5. *Quercus rugosa* with *Tilandsia* in the canopy (photograph by A. Chimal)

3.2.5   Mountain cloud forest

The mountain cloud forest (Leopold, 1950) is a very rare and beautiful vegetation type, which is present mainly on the west slope of the Iztaccíhuatl volcano in a much-reduced area. The species present in this forest require a very high degree of humidity and therefore, they are found exclusively in the deepest part of some cañadas between the 2,500 and 2,900 m. These sites are always very well protected from sun and strong winds keeping an annual mean temperature of 12 to 14° C; they receive an annual rainfall above 1,000 mm. The soils are deep, rich in organic matter and humid almost all year round.

The tree layer goes from 10 to 25 meters high and it is very dense. Most of the species are perennial and the tree canopy hardly shows any changes in greenness during the change of seasons. Depending on the location and the degree of disturbance, the association of the dominant species changes. Among the most characteristic elements of this type of forest, we find *Cornus excelsa*, *Ceanothus* and *Rubus*.

One of the best-preserved areas of this type of forest is the deepest parts of the cañadas of Ococintla and Tlapexpa (Figure 2.6). The top tree layer is formed by mature *Pinus montezumae* trees (35 to 45 meters high) associated on the lower tree layer with a dense canopy (90%) of *Prunus brachybotrya* and *Garrya laurifolia*. The shrub layer is also dense mostly with *Eupatorum pazcuarense*. Adjacent areas have isolated individuals of the following species:
Clethra mexicana, *Cornus disciflora*, *Ilex tolucana*, *Meliosma dentata*, *Quercus laurina*, *Abies religiosa*, *Alnus arguta*, *Buddleia cordata*, *Cornus excelsa*, *Cupressus lindleyi*, *C. lusitanica*, *Ageratina mairetianum*, *Pinus ayacahuite*, *P. patula*, *P. pseudostrobus*, *Prunus serotina* ssp. *capuli*, *Quercus rugosa*, *Sambucus mexicana*, *Oreopanax xalapensis*, and *Viburnum stenocalyx*.

Among the most common shrub species found in this type of vegetation are: *Archibaccharis sescenticeps*, *Cestrum terminale*, *Iresine ajuscana*, *Laumourouxia xalapensis*, *Lippia umbellata*, *Montanoa frutescens*, *Urtica dioica*, *Solanum apendiculatum*, *Salvia mocinoi* and *Salvia gesneriflora*. *Ageratina isolepis*, also present in fir and oak forests is reported by Rzedowski and Rzedowski (1979) as a species endemic to the Valley of Mexico whereas *Ageratina ramireziorum*, another endemic, is found also in oak forests.

Figure 2.6. Cloud forest (photograph by A. Chimal)

The most common herb species in these communities are: *Bidens ostruthioides, Bromus dolichocarpus* and *Peperomia hispidula*. In addition, also ferns are present: *Adiantum andicola, Dryopteris wallichiana* Subsp. *Wallichiana, Phanerophlebia nobilis* and *Pteris cretica*. Other interesting vegetation forms present in these communities are the woody climbing species: *Archibaccharis hirtella, Celastrus pringlei, Clematis dioica, Philadelphus mexicanus, Smilax moranensis, Solanum appendiculatum* and *Valeriana clematis*. Finally, we also *Tillandsia* and *Peperomia* can be found.

3.2.6   Mixed forests

When they are mixed, these forests are named according to the dominant species and this varies according to the altitude as well as changes in the local conditions of humidity, exposition and soil type. In this way, we observe that the fir-pine forests are located in most of the transitional areas, as fir becomes less dense at the open parts of the cañadas. In most cases, *P. montezumae* is the codominant pine species but also *P. ayacahuite* and *P. hartwegii*; the latter in the higher parts of the mountains (Figure 2.7).

Figure 2.7. Mixed forest (photograph by A. Chimal)

When the transition form the pure fir forests in the cañadas is located in areas lower than 2,800 m and towards protected and more humid areas, the co-dominant species can be *Cupressus* spp., *Quercus* spp., *Garrya laurifolia* and *Buddleia cordata*. These mixed communities can have as associated shrub species *Baccharis conferta* and *Senecio* spp. In the herb layer it is common to find *Acourtia hebeclada* and *Halenia brevicornis*.

The other very important mixed forest is the one formed by pine and oak trees. These forests are usually combinations of *P. montezumae* or *P.leiophilla* with *Quercus rugosa*, *Alnus jorullensis* ssp. *jorullense* and *Arbutus xalapensis* in the tree layer. The dominant shrub species are *Salvia* spp. and *Fuschia microphylla*; in the herb layer *Conopholis alpina* is dominant. The canopy cover of the tree layers is mostly high, ranging from 50 to 80 % in the high tree layer and mostly 50% in the lower tree layer. The canopy cover of the shrub layer is also quite high, ranging from 35 to

50% and the herb layer much lower around 20%. There is usually a ground layer with an average of 10% cover. In some areas at the lower limit of the altitudinal range covered by this research (2500 m), *Pinus montezumae* can be found combined with very short individuals of *Quercus frutex* (about 1 m tall). These communities are found in very shallow soils with annual rainfall of 700 to 900 mm and mean temperatures from 9 to 13 °C. The slopes are very steep (40%) and associated shrub and herb species are very few: *Baccharis conferta* in the shrub layer and *Brickelia* sp. and *Castilleja* sp. in the herb layer. Rzedowski and Rzedowski (1979) report that these types of communities are strongly induced by frequent fires.

3.2.7   Grasslands

Alpine Grasslands

These communities are given many different names: high paramo (Miranda, 1947), alpine prairie (Beaman, 1962) or *zacatonal*, a word derived from the aztec word ***zacate***, which refers to those high tall and clumped grasses such as *Muhlenbergia* and *Festuca*. We will use the term of Rzedowski (1978): alpine grassland although it is not very clear why high mountains located in America should borrow their name from the celtic voice *alp* which identifies the European high mountains.

This vegetation type is endemic of the central mountains of Mexico and it is located in the Iztaccíhuatl and Popocatépetl volcanoes between 3,800 m where *P. hartwegii* almost ends till the 4,500 m where all vegetation stops growing shift into the volcanoes rocks, bare soils and glaciers. The soil under these grasslands is mostly sandy derived from volcanic rocks. The climate is extremely cold with temperatures from 3 to 5 °C all over the year and rainfall from 600 to 800 mm (Figure 2.8).

The most relevant plant species is *Muhlenbergia quadridentata* and according to other authors as Beaman 1965, Cruz 1969 and Almeida-Leñero *et al.* 1997a,b, three associations can be distinguished: between 3,700 and 3,800 m *M. quadridentata* is the dominant species.

Figure 2.8. Alpine grassland (photograph by R. Bobbink)

From 3,800 to 4,200 m this species is associated with *Calamagrostis tolucensis* and *Festuca tolucencis*. Finally, at the upper part from 4200 to 4500 m, *M. quadridentata* is associated with *Festuca livida* and *Arenaria bryoides*; the latter forming thick patches which is a characteristic adaptation to the extreme cold. Other species that are commonly found with these associations are *Poa conglomerata*, *Trisetum spicatum*, *Senecio orizabensis*, *S. procumbens* and *Draba jorullensis*. In very moist or even swampy places, *Carex* spp. is present and on the rocks and humid soils several moss species are common.

Induced Grasslands

Although there is usually a strong discussion as to when grassland is induced and when it is natural, the characteristic species of the alpine grasslands are usually very reduced in their density in disturbed areas. As a result, we can find patches of grasslands in the volcanoes area with the following grasses and herbs:

Muhlenbergia macroura, M. quadridentata, Festuca tolucensis, Calamagrostis tolucensis, Stipa ichu, Cinna poeformis, Deschampsia pringlei, Cyperus sesleriodes, Vaccinum geminiflorum, Gentiana ovatiloba, Phacelia platycarpa, Pernettya ciliata, Arenaria bourgaei, Stellaria cuspidata.

In places protected by rocks, we have *Heuchera orizabensis* and *Ageratina* spp.; quite often the surface of the rocks is covered by *Villadia batesii.*

3.2.8   Secondary vegetation

An unfortunately increasing vegetation type is the secondary vegetation that is taking over many pine and pine-oak forests as a consequence of human activities (Figure 2.9). When this kind of vegetation is derived from fir, oak or pine forests it is common to find several *Senecio* species dominating. As we go south, we also find *Coriaria ruscifolia, Reseda luteola* and *Brassica campestris.*

When this secondary vegetation is close to former cultivated areas, we can find the following species:

Argemone platyceras, Eruca sativa, Castilleja spp., *Geranium seemanni, Lupinus camprestris, Phacelia platycarpa, Ranunculus petiolaris* var. *arsenei, Reseda luteola, Senecio salignus, Zephyranthes carinata.*

Figure 2.9. Induced grassland by burning and consequetively cattle grazing (photograph by
A. Chimal)

4.   DISCUSSION

The vegetation characterization in this Chapter constitutes a quite
complete and updated description of the forests in the Iztaccíhuatl and
Popocatépetl volcanoes. The species list accounts for a most extensive
fieldwork during 4 years and it will certainly provide a useful reference to
future investigations in the area. However, the use of remotely sensed
images to produce a supervised classification requires still of further field
validation and more detailed sampling along several gradients.

The importance of improving this classification relies on the
possibilities that spatial analysis provides for a better understanding of the
ecological relationships in the region. To show this, a series of graphs were
elaborated to analyse the vegetation distribution from the supervised
classification shown in Figures 2.10. 2.11 and 2.12 with respect to altitude,
temperature, slope in degrees, and aspect/orientation.

Figure 2.10 shows a clear separation of some vegetation types along the altitudinal gradient. One of the most conspicuous vegetation types to this respect is *Pinus hartwegii* due to its marked separation from the rest types within the range of 3,500 to 4,000 m. The left part of the *P. hartwegii* distribution line, below 3,100 m belongs to the open conifer forests, some of which might also comprise areas with shrubs and fruit trees.

With much lower density values, the alpine grasslands also separate themselves clearly from 3,600 to 4,400 m. Higher than that, the values from the graph represent overestimated grasslands on the bare tops of the volcanoes, probably due to the higher reflection values of small clouds and bare rock. The grasses on lower areas show their highest abundance between 2,500 and 2,800 m with a considerable reduction from 2,800 to 3,600 m, which is the altitudinal range most tolerated by Abies, *religiosa*, *Pinus montezumae* and *Cupressus* spp. forests.

Mixed conifer forests, broad-leaved forests, dense conifer, and mixed conifer/broad-leafed forests have a wider distribution range with no particular preference. The higher values for broad-leaved forests between 3,200 and 3,400 m very probably belong to dense conifer forests with *Abies religiosa* and mixed conifer/broad-leafed forests. Something to be cautious about are the values above 3,700 m, which should correspond only to *Pinus hartwegii* and alpine grasslands.

As for the relationship between temperature and vegetation, a climatic zone map has been used that separates eight temperature ranges (Flores, 1997). This climatic zone map includes both temperature and precipitation data and it also accounts for regional variations in topography and altitudinal ranges. Taking this into account, Figure 2.11 shows that *Abies religiosa* is quite akin in its distribution to broad-leaved forests, this is to be expected due to their similar preferences for more humid and temperature conditions.

Grasses separate in what can be interpreted as the difference between alpine and induced grassland, having the former with highest values between -1.5 and 16 °C. With a similar pattern but with higher values and a more restricted distribution (1 to 16 °C) we find *P. hartwegii*.

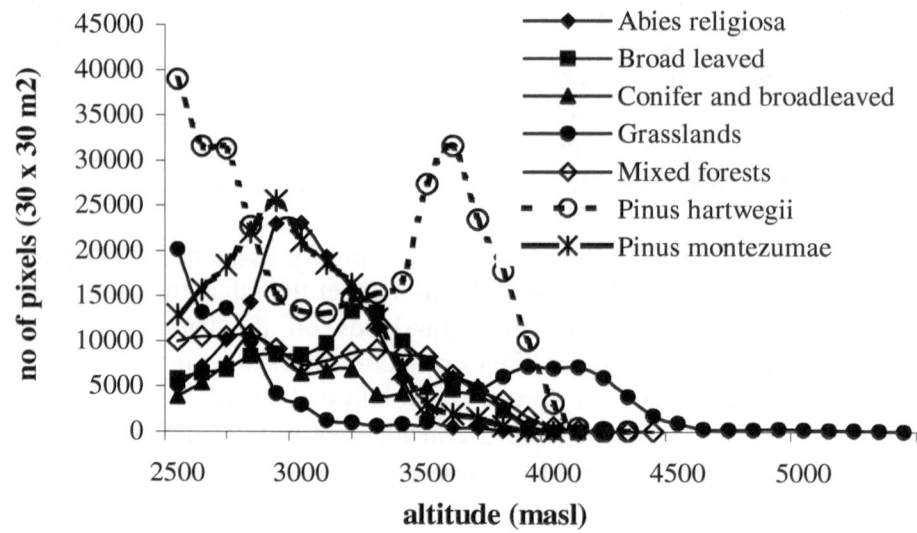

Figure 2.10. Relation between altitude (m) and vegetation

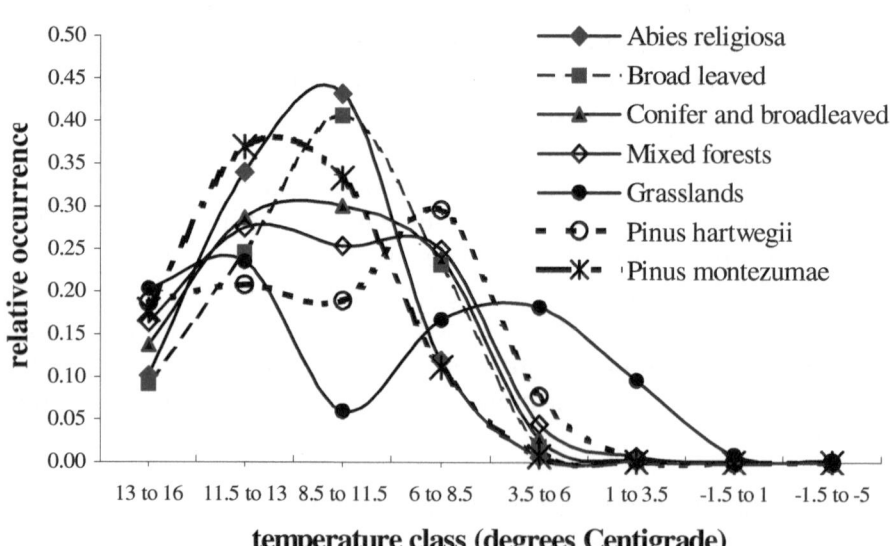

Figure 2.11. Relation between temperature (^{0}C) and vegetation

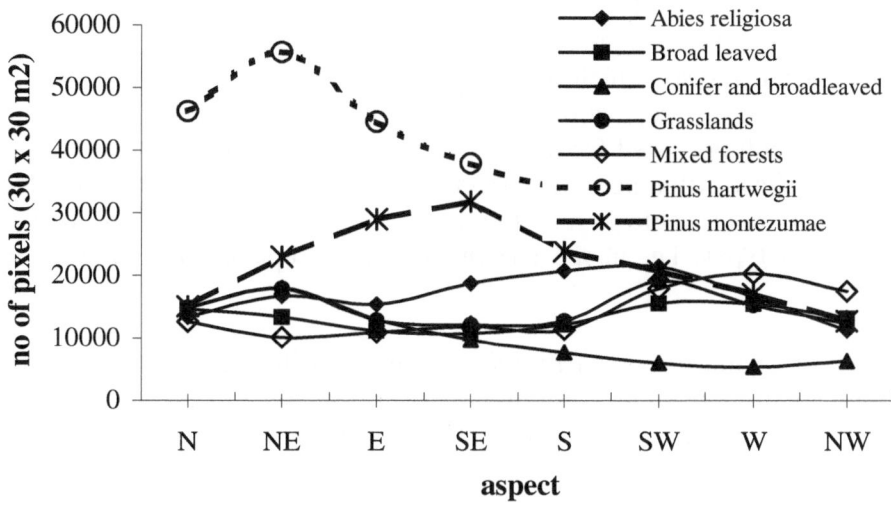

Figure 2.12. Relation between orientation and vegetation

As we could see in the relationship with altitude, open conifer forests and induced grasslands correspond, with respect to temperature, to warmer areas with values ranging from 8.5 to 13 °C. A final note with respect to the temperature gradient is that *Pinus montezumae* shows a more clear separation from *Abies religiosa* than it did with respect to altitude having a slightly higher preference for temperatures of 8.5 to 11.5 °C whereas the preference of fir forests is markedly towards areas with temperatures ranging from 6 to 8.5 °C.

Exposition conditions are also very important in determining the distribution of the different vegetation types. An alternative way to see the effect of the exposition in the vegetation is provided by aspect, which is a simplification of the slope orientation. These structural changes in the terrain have a strong effect on local climate conditions, which in some cases can be reflected in the vegetation composition. In general, the Iztaccíhuatl and Popocatépetl mountains are separated basically towards two big areas; east and west, with a very heterogeneous area on the north; the aspect relationship shows the consequences of this.

The spatial analysis of 8 aspect classes in relation to vegetation (Figure 2.12) shows only a significant separation of *Pinus hartwegii* and *Pinus montezumae* forests which might be in reality a consequence of the high abundance of this vegetation type with respect to the others. It is possible that the resolution used for the classification (pixel size of 30 x 30 meters) does not show differences that affect vegetation distribution at a more local scale. It is remarkable to note the overlapping distribution range for these vegetation types, i.e. in the terrain they truly separate due to slope and aspect.

On one hand, *Pinus hartwegii* is mainly distributed in the higher parts, which are more open and present larger surfaces towards the northeast. The open conifer forests are also more characteristic of the eastern area towards the estate of Puebla.

It is remarkable to note the distribution of the dense conifer and mixed conifer/broad-leafed forests in their preference for the north, north-east and east orientations. These occur basically on the northern part of the mountains where the hill-lands region favour the environmental conditions preferred by this vegetation types.

With an opposite distribution, we find a relative higher abundance in the other range of the gradient (southwest, west and northwest) the mixed conifer forests, the broad-leaved forests, and the alpine and induced grasslands. *Abies religiosa*, on the other hand, also follows this pattern but with a more wide range covering also with high abudance at locations oriented southeast and south.

ACKNOWLEDGEMENTS

We would like to thank the following persons who contributed significantly with their work to the elaboration of this Chapter: Miriam Díaz, Marisa Silva, Laura Gómez Aíza and Gilberto Hernández.

REFERENCES

Almeida-Leñero, L., Cleef, A.M., Herrera, A., Velazquez, A. and Luna, I. (1994). El zacatonal alpino del Volcán Popocatépetl, México, y su posición en las montañas tropicales de América. Phytocoenologia 22 (3) 391-436.

Almeida-Leñero, L., Cleef, A.M. and González, A. (1997a). Fitosociología de la vegetación alpina de los volcanes Popocatépetl y Nevado de Toluca, región central de México. In: Vegetación, Fitogeografía y Paleoecología del Zacatonal Alpino y Bosques Montanos de la Región Central de México (Almeyda-Leñero, L.). pp.61-88.Tesis de Doctorado. Universidad de Amsterdam.

Almeida-Leñero, L., Cleef, A.M. and Velazquez, A. (1997b). Fitosociología del bosque de coníferas del volcán Popocatépetl, México. In: Vegetación, Fitogeografía y Paleoecología del Zacatonal Alpino y Bosques Montanos de la Región Central de México (Almeyda-Leñero, L.). pp.61-88.Tesis de Doctorado. Universidad de Amsterdam.

Anaya, A.L., Hernández, R. and Madrigal, X. (1980). La vegetación y los suelos de un transecto altitudinal del declive occidental del Iztaccíhuatl (México). Boletín Técnico No. 65. Instituto Nacional de Investigaciones Forestales.

Beaman, J.H. (1962). The timberlines of Iztaccíhuatl and Popocatépetl, México. Ecology 43: 377-385.

Beaman, J.H. (1965). A preliminary ecological study of the alpine flora of Popocatépetl and Iztaccíhuatl. Boletín de la Sociedad Botánica de México.

Cárdenas, A. (1987). Nuevos registros para la flora de musgos de México y del Valle de México. Anales Inst. Biol. UNAM 58, Ser. Bot.: 93-96.

Chávez, J.M. and Trigo, N. (eds.) (1996). Programa de Manejo para el Parque Nacional Iztaccíhuatl-Popocatépetl. Colección Ecología y Planeación, Universidad Autónoma Metropolitana, Xochimilco, México. 273 pp. 12 mapas.

Cruz, R. (1969) Contribución alconocimiento de la ecología de los pastizales en el VLLE DEmÉXICO. Tesis.Escuela Nacional de Ciencias Biológicas. México. 235 pp.

Eastman, J.R. (1993). *IDRISI, version 4.1, update manual.* Clark University Graduate School of Geography, Worcester, MA. 207 pp.

Ern, H. (1972). Estudio de la vegetación en la parte oriental del México Central. Comunicaciones del Proyecto Puebla-Tlaxcala 6: 1-6.

Escamilla, M.E. (1996). La vegetación alpina y subalpina del declive occidental del volcán Popocatépetl, México. Tesis de licenciatura. Facultad de Ciencias, UNAM, México, 50 pp.

Flores, M. (1997). Diseño y aplicación de una metodología para la identificación de ecotonos en el Parque Nacional Iztaccíhuatl-Popocatépetl y su área de influencia. Tesis de Licenciatura. Departamento el Hombre y su Ambiente, UAM-Xochimilco, México. 60 pp. + 14 mapas.

Fuentes, A.L. (1975). El paisaje en el piedemonte poblano de los volcanes Popocatépetl e Iztaccíhuatl. Boletín No. 6, Insitituo de Geografía, UNAM.

Guzmán, G. (1966). Hongos (Macromicetos) comunes en la ruta Amecameca-Tlamacas (Volcán Popocatépetl), México. En: Guía de las excursiones del Tercer Congreso Mexicano de Botánica, Sociedad Botánica de México, México; pp. 13-16.

Guzmán, G. (1972). Algunos macromicetos, líquenes y mixomicetos importantes en la zona del volcán Popocatépetl. En: Guías botánicas de excursiones en México, I Congreso Latinoamericano y V Mexicano de Botánica, Sociedad Botánica de México, México; pp. 17-43.

Leopold, A.S. (1950). Vegetation zones of Mexico. Ecology 31: 507-518.

Mabberley, D.J. (1998). The Plant Book. A portable dicitonary of the vascular plants. University Press. Cambridge. 858 pp.

Madrigal, S. X. (1967). Contribución al conocimiento de la ecología de los bosques de oyamel [*Abies religiosa* (H.B.K.) Schl. et Cham.] en el Valle de México. Instituto Nacional de Investigaciones Forestales,. Boletín Técnico 18. México, D. F. , 94 pp.

May-Nah, A. (1971). Estudio fitoecológico del Campo Experimental San Juan Tetla, Puebla, México. Tesis. Escuela Nacional de Ciencias Biológicas, México, d.F. 130 pp.

Miranda, F. (1947). Estudios sobre la vegetación de México - V. Rasgos de la vegetación en la cuenca del Río de las Balsas. Revista de la Sociedad Mexicana de Historia Natural 8 (1-4): 95-114 + 15 láminas.

Obieta, M.C. and Sarukhan, J. (1981). Estructura y composición herbácea de un bosque uniespecífico de *Pinus hartwegii*. I. Estructura y composición florística. Boletín de la Sociedad Botánica 41: 75-125

Rzedowski, J. (1970). Nota sobre el bosque mesófilo de montaña en el Valle de México. Anales de la Escuela Nacional de Ciencias Biológicas. México, 18: 91-106.

Rzedowski, J. (1975). Flora y vegetación de la Cuenca de México. En: Memorias de las obras del Sistema de Drenaje Profundo del Distrito Federal, Tomo 1, Departamento del Distrito Federal, México; pp 79-134.

Rzedowski, J. (1978). Vegetación de México. Limusa. México D.F. , 432 pp.

Rzedowski, J. and Rzedowski, G.C. (Eds.) (1979). Flora Fanerogámica del Valle de México. Vol. I Ed. CECSA. México, D.F., 403 pp.

Rzedowski, J. and Rzedowski, G.C. (Eds.) (1985). Flora Fanerogámica del Valle de México. Vol. II. Ed. ENCB, IPN e Instituto de Ecología A.C. México. 674 pp.

Chapter 3

Man-induced changes in vegetation cover in the Iztaccíhuatl-Popocatepetl Region

Roland Bobbink*, Gerrit W. Heil** and Betty Verduyn**

*Landscape Ecology,** Plant Ecology, Faculty of Biology, Utrecht University, The Netherlands*

Key words: Erosion, Deforestation, Fragmentation, Land use changes, Ecological threats, RS, vegetation.

Abstract: An analysis has been carried out on the changes in natural communities in the Iztaccíhuatl-Popocatépetl region in the period 1986 to 1997 using satellite images and vegetation ground data. The ecological impacts of these recent changes are discussed in relation to the increase in human pressure and future hazards. The most obvious changes were observed in the forested parts (2,500 – 4,000 m) of the area. More sparse *Pinus hartwegii* stands were found at the highest forest sites in 1997 than in 1986; secondly, dense *Pinus*, *Abies* or mixed forest decreased in density at the lower slopes. Finally, a large increase in area with shrubs and very sparse tree cover was found in the central-eastern slopes over a large altitudinal gradient. Via cross comparison it became evident that total forest density decreased with 13.2 %, and forest fragmentation increased, too. A detailed cross-comparison showed that the observed deforestation rate is strongly influenced by its accessibility (distance to trails and its steepness). Three main causes have been identified for this decline in forest cover, viz. (a) enhanced frequency of burning, associated with more livestock grazing; (b) more destructive forestry activities, and (c) increased settlements, especially at the eastern part of the region. A considerable part of this high volcanoes region has an increased risk of erosion because of this deforestation. In addition, non-fossil CO_2 emission will strongly increase by the deforestation process. Finally, this gradual degradation may lead to a decline in typical, partly endemic biodiversity in future. It is concluded that adequate natural resource management is highly needed in the near future.

G.W. Heil et al. (eds.), Ecology and Man in Mexico's Central Volcanoes Area, 49–71.

1.   INTRODUCTION

The Iztaccíhuatl-Popocatépetl mountainous area (abbreviated as Iztac-Popo area) is in the central part of the Transversal Neovolcanic Axis of Central Mexico. The area is only 50 km to the east of one of the largest cities of the world (Mexico city), and ca. just west of the relatively densely populated state of Puebla. Because of its large altitudinal gradient (from 2,400 – 5,565 m) a range of typical vegetation belts are found in this area. The Iztac-Popo area is already lengthy known for its high alpine bunch grasslands ("zacatonal") and the typical coniferous and mixed forests (e.g. Rzedowski 1978, Chávez and Trigo 1996, Velázquez *et al.* 2000). Both forests and grasslands are of major importance for the biodiversity of plant and animals in this mountain region of Central Mexico (Chapter 1). Parts of the Iztac-Popo region became already a forest reserve in 1929, and the volcanoes and their surroundings above 3,000 m were declared a national park in 1935, covering almost 60,000 ha. The limits of the Iztac-Popo national park were raised to 3,600 m in 1948 (37,350 ha) to facilitate forest exploitation in the northwestern parts of the area (Chávez and Trigo 1996).

As in many developing countries, the urban population and its areas strongly increased in the last half of the 20th century. This certainly holds for the Valley of Mexico; the population of the metropolitan area of Mexico City was just above 1 billion in 1940, 2.1 in 1960, but already 12.5 in 1980 and probably around 20-25 million in the 1990s (Figure 3.1, Chapter 1; e.g. Rees 1996). Moreover, the total surface of the urban settlements increased enormously in this period, and the area occupied with agriculture did certainly not decrease in this period. In addition, the human settlements also spread over the whole region of the Valley of Mexico, invading the foothills of the Iztac-Popo area more and more. In addition, the population pressure increased in recent decades in the Puebla parts of the area, too (López-Paniagua *et al.* 1996).

It is thus likely that these changes in population pressure in the last decades of the previous century have lead to changes in land use in the volcanoes area. Especially, an intensivation of the use of the natural vegetation communities of the Iztac-Popo area can be expected. Instead of small-scale wood logging for local use, more intensive logging with modern machinery and a transition from traditional pastoral systems to a more intense system with higher frequencies of burning and higher grazing pressure, especially by cattle and sheep, are probably most important in this respect (Chávez and Trigo 1996; Velázquez *et al.* 2000). In the present chapter an analysis will be given of the changes in natural communities in the Iztac-Popo area in the period 1986 to 1997 using satellite images and ground data of the vegetation. Especially the changes in the density of the different forested areas are treated in detail. For details on the floristic composition of the different communities, see Chapter 2. In addition, the changes in land cover and the pattern of these changes are related to the observed changes in land use and the increase in disturbances in the area. The ecological consequences of these disturbances for the structure and function of the natural (forest) systems are dealt with, including the possible impacts on erosion. Finally, the risks for hazards in the future are shortly discussed.

2. CHANGES IN VEGETATION COVER

2.1 Methodology

Remotely sensed data in combination with field data were used to analyse the spatial distribution of the vegetation, and the recent changes in its distribution over an 11-year period (Figure 3.2).

The changes in vegetation cover have been determined with two RS-images taken from the Iztac-Popo area in the dry season; (1) a Landsat-MSS image from January 1986, and (2) a Landsat-TM image from February 1997. The MSS image had a ground resolution of 80 x 80 m, and consisted of four wavelength bands between 0.5 and 1.1 μm. The TM image had 7 bands from 0.45 to 2.35 μm with a resolution of 28.5 x 28.5 m. In our analysis we used the 60 x 30 km central part of the original images. A TM image of January 1991 was originally incorporated in our analysis, but the available image had unfortunately too high cloud cover to be utilized in the analysis of vegetation change.

The analyses were done with a GIS system (IDRISI for Windows, v. 2.0). Both RS-images were resampled to a resolution of 50 x 50 m to facilitate comparison between both images, and with the existing maps of the area (e.g. Digital Terrain Model, road map, hydrology map). The MSS image of January 1986 was georeferenced, whereas the TM image was firstly radiometrically corrected and then geometrically corrected with the use of the fully georeferenced TM image of 1991. The classification of the vegetation cover of both RS-images was carried out with supervised classification, a method which generally yields better description of (tropical) forest than unsupervised classification (Eastman 1993, Lilesand and Kiefer 1994).

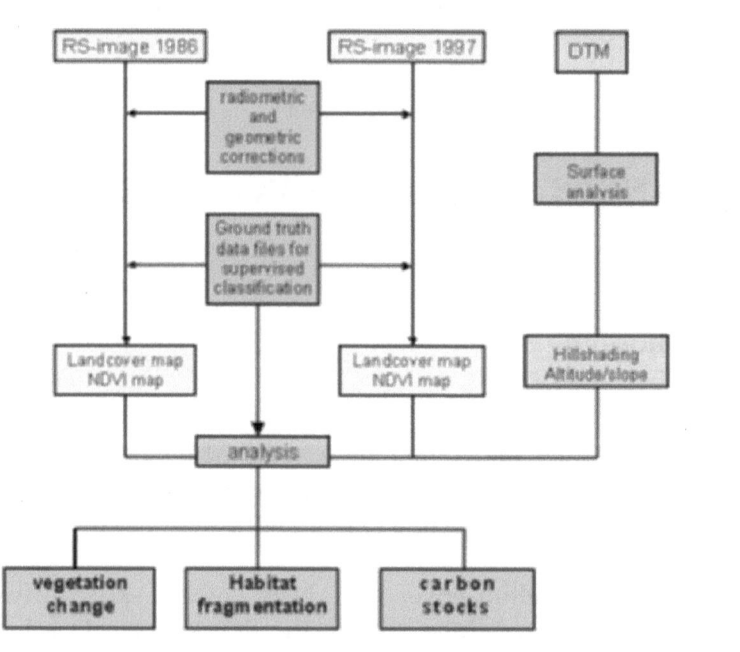

Figure 3.1. Scheme of the sampling and data processing methodology.

In this analysis, we used a set of training pixels (n=144) to derive the spectral signature of the different vegetation and land cover classes (Table 3.1). The information on the training pixels were collected spread over the Iztac-Popo area between 1993 and 1997, using a GPS system to locate the site and a detailed description of the vegetation layers or the land cover. For the different forest units only training pixels with a quantified cover percentages of the tree layer(s) were used in this procedure. The remaining locations without cover estimates of the tree layers were utilized to validate the results of the supervised classification. The information on the forbs and annual grass understory of the forests was not used, because both RS-images were made in the dry season. The field data of pure broad-leaved stands (respectively *Quercus* and *Alnus* species; rare at low to intermediate altitudes) and of secondary grasslands were too few to be used in the classification. The supervised classification was done with a maximum likelihood procedure, utilizing the green, red, near infrared (NIR) bands and the normalized difference vegetation index (NDVI) composite (NDVI = (NIR − R) / NIR + R).

Table 3.1. The distribution of the training pixels, used in the supervised classification, over the different vegetation and land cover classes

Land use type	No.
Volcano peaks	
Snow and ice	n=7
Bare rocks	n=11
More or less natural vegetation	
Alpine grasslands ('zacatonal')	n=17
Abies religosa- forests	n=25
Pinus spp. forests (with P. hartwegii, P. montezumae & P. leiophylla)	n=24
Open Pinus hartwegii stands (5-15 % tree cover)	n=9
Mixed forests (coniferous and broad-leaved tree species)	n=14
Degraded areas	
Shrubs (sites with shrubs and very low Pinus cover (< 5%))	n=7
Man-made landscapes	
Cropland (most bare in this season)	n=18
Urban areas	n=12
Total number of training pixels	n=144

Finally, a ground thruthing of the supervised classification was carried out at 7 altitudinal gradients spread over the forest areas of the Iztac-Popo region in September 1998. The agreement between the classified and the present forest or land use types proved to be consistent.

2.2   Changes in vegetation cover from 1986 to 1997

The Iztac-Popo area is very important for biodiversity in the coniferous and mixed forest belt of the mountainous areas of central Mexico, besides its presence of alpine grasslands (e.g. Challenger 1998). The distribution of the vegetation cover in 1986 can be summarized as follows. One percent of the area above 2,500 m (total of approximately10 km^2) was covered by snow and rocks, forming the highest parts of the two volcanoes, whereas alpine bunch grasslands (> ± 3900 m) occupied almost 17 % (Table 3.2).

Table 3.2. Percentage of the area above 2500 m occupied by the different vegetation and land cover types.

Land use type	1986	1997
Volcano peaks		
Snow and ice	0.04	0.15
Bare rocks	0.77	0.83
More or less natural vegetation		
Alpine grasslands	16.6	16.0
Abies religosa forests	8.6	7.9
Pinus spp. forests	13.0	18.3
Open Pinus hartwegii stands	13.2	4.3
Mixed forests (coniferous/broad-leaved)	18.8	16.0
Degraded areas		
Shrubs with very sparse Pinus vegetation	0.98	12.6
Man-made landscapes		
Cropland	18.6	8.7
Urban areas	9.5	15.2

Figure 3.2. An overview of a typical *Abies religiosa* forest in a steep valley (photograph by R. Bobbink)

Dense *Abies* forest was clearly more present at intermediate altitudes at the western slopes of the Iztac-Popo area, but also found in steep valleys at the eastern side (Fig. 3.2). Totally, it engrossed ca. 9 % of the area. The highest parts of the slopes with forests (ca. 3,200 – 3,900 m) were formed by open *Pinus hartwegii* stands with large tussock grasses in between (Fig. 3.3), forming 13 % of the area above 2,500 m. In the northern and central-eastern parts of the Izta-Popo area, these stands were relatively more extensive than on the western slopes. Mixed coniferous/broad-leaved stands (18 %) and forests with different *Pinus* species (13 %) were present in the lower parts (2,600 – 3,300 m) of the forested areas, especially in the northeastern and southeastern parts. It can be concluded that ca. 71 % of the Iztac-Popo area above 2,500 m was in 1986 covered by more or less natural vegetation or rocks and snow, with almost 54 % of the area with forests. After analysing the TM image of 1997, a number of changes became obvious, although the general pattern of the vegetation cover as found in 1986 was still present (Table 3.2).

Figure 3.3. Open *Pinus hartwegii* forest at ca. 3600 m in the Iztac-Popo area (photograph by R. Bobbink).

The distribution at the highest parts of the volcanoes, including the alpine grasslands (Fig. 3.4), hardly changed in the 11-year period. Because of the volcanic activities of the Popopatépetl since December 1994, some recent changes in vegetation or land cover at high altitudes can be observed, when comparing both satellite images. The cover of alpine grassland and open *P. hartwegii* forest seemed to be replaced by bare soil over an altitudinal range of 500 m in the northeastern slope of the Popocatépetl (ca. 3 x 2 km). This is probably caused by mud or lava flows, which resulted from the high volcanic activity in the last three years of the observational period. In the man-made landscapes at the foot of volcanoes, an increase in urban areas compared with the cropland area was seen.

The most obvious changes between 1986 and 1997 were observed in the forested parts of the Iztac-Popo area. The following trends were most evident:

- markedly more very sparse *Pinus hartwegii* stands were found at the highest forest sites in 1997 than in 1986;
- at the lower slopes near the agricultural land use units, the area with dense *Pinus*, *Abies* or mixed forest decreased;

- a large increase in shrub area with very sparse tree cover was found in the central-eastern slopes over a large altitudinal gradient.

The quantification of deforestation or afforestation was, in addition, analysed in detail by combining the 10 classes (Table 3.2) to three forest classes and two non-forest classes. The following forest classes were formed, based upon tree density: (1) dense forest (*Abies* & mixed forest); (2) medium dense forest (*Pinus* spp.-stands); and (3) open forest (open *Pinus hartwegii* stands). Besides, two non-forest classes were used; (4) shrub with very sparse tree cover (*Pinus* or broad-leaved), and (5) low vegetation or 'bare' (alpine bunch grasslands, rocks, cropland and urban). Via cross comparison of the 1986 and the 1997 maps it became evident that in the investigated 11-year period the total forested area (types 1-3) decreased with 13.2 %, compared to the forested area in 1986, mainly due to a decline in dense and open forest stands (Table 3.3).

Table 3.3. Changes (% of the area above 2,500 m) in forest density between 1986 and 1997 in the Iztac-Popo area. For grouping of classes, see text.

Year	Dense forest	Medium forest	Open forest	Shrub	Bare	Total
1986	27.4	13.0	13.2	1.0	45.5	53.6
1997	23.9	18.3	4.3	12.6	40.9	46.4

This deforestation rate (1.2 % per year) is in line with the overall annual deforestation rates in Mexico (0.55 – 1.30 %; Harcourt and Sayer 1995; Challenger 1998). Locations with shrubs, or very sparse *Pinus* vegetation have increased remarkably in this 11-year period in the Iztac-Popo area. An overview of all transitions between the forest and non-forest classes is additionly given on the accompanying CD. The net decrease in total forested area (Table 3.3) is based upon both increases and decreases in forest cover. At some locations stands with increased tree densities were observed, but the 'deforestation' trend was significantly found at much more locations. The spatial distribution of deforestation trends is also presented in a simplified form in Figure 3.5.

After analysing the TM image of 1977, a number of changes became obvious, although the general altitudinal pattern of the vegetation cover as found in 1986 was still present (Table 3.2). It showed that deforestation especially occurred at three parts of the Izta-Popo area. The higher parts of the volcanoes, the foothill at low altitudes and the central eastern part showed a remarkable shift to lower forest densities ('deforestation').

Figure 3.4. Alpine bunchgrasslands ("zacatonal") above the treeline in front of the Iztaccíhuatl (photograph by R. Bobbink).

The steep middle parts of the slope, especially in the western site, were relatively unaffected in this research period. Based upon the NDVI maps of 1986 and 1997, and the relationship between above-ground biomass and NDVI, it was estimated that approximately 400,000 Mg of carbon disappeared from the forests in these 11 years (see Chapter 7). This can be considered as a relevant contribution to the non-fossil CO_2 emissions of Central Mexico.

Besides the reduction in area occupied by forests, fragmentation of the stands is also an ecologically important phenomenon (e.g. Hobbs 1998). Fragmentation of the forests and the other vegetation types was analysed with the nearest neighbour method (Idrisi manual; Eastman 1993). In this method, the vegetation class of a pixel is compared with its 8 surrounding pixels, giving a value for the existing fragmentation. This procedure was carried out for both the TM images; the increases or decreases in fragmentation between 1986 and 1997 were determined by cross comparison of the fragmentation maps.

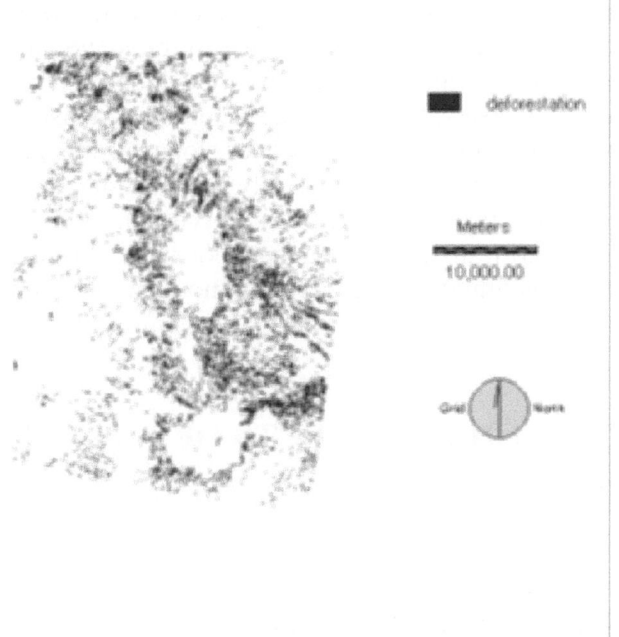

Figure 3.5. Pattern of deforestation above 2,500 m from 1986 to 1997. All transitions in the direction of lower forest density in 1997 were taken together (see text).

The results clearly showed that at many locations the forested areas were seriously more fragmented in 1997 than in 1986, in a few locations fragmentation also decreased. This could be caused by regrowth in gaps of forested areas, but also by the formation of larger areas with a single, mostly degraded, vegetation type (e.g. shrubs or very sparse *Pinus*).

2.3 Accessibility and changes in forest cover

To get a better insight of the pattern of the changes in vegetation cover, a 10.5 x 10.5 km square between 21° 37'N, 52°72' E and 21° 33' N, 53° 35' E in the northern part of the Iztac-Popo is overlaid, both in 1986 and 1997, with the digitised road map of the region. Interpreting the changes in forest cover from 1986 to 1997, a distinct increase in very open *Pinus* stands or shrubs became prominent, as shown before in the maps of the whole area. It is, however, evident that the rate of deforestation is not randomly distributed over this high mountainous region (see also Figures on CD), but highly related with the distance to trails and small roads. The decrease in forest density in the 11-year period is considerably higher aside these trails and small roads, both unpaved. Moreover, dense *Abies* or *Pinus* forest stands were found only on the steeper slopes of this northern section of the Iztac-Popo area in 1997. This clearly demonstrates that the observed deforestation is strongly influenced by the accessibility of a forest stand; especially distance to trails for the transport of wood and the steepness of the stand are the most important in this rural mountainous landscape.

3.   ECOLOGICAL EFFECTS OF CHANGED LAND USE

3.1   Traditional use of the forests

The natural resources of the forests in the Iztac-Popo areas are, of old, used by the people living in its surroundings, as in other landscapes of temperate Mexico. Indigenous people, Spanish colonisers and the present mixed population all have impacted the natural communities of Central Mexico (e.g. Challenger 1998; Velázquez *et al.* 2000). The utilization pressure on the natural ecosystems was, however, much lower than in present time. Not much is known of the degree of use of the area in pre-Hispanic times, but 17 archaeological sites are discovered in the Iztac-Popo region, especially at the higher slopes (Chavez and Trigo 1996). These sites were probably related with the special interest of the different pre-Hispanic cultures for these high volcanoes. Different cultures (Aztec, Otomiés, Nahuas and Chichimecas) were established in the near surroundings of the volcanoes. It is quite certain that sulphuric material and some other minerals were collected at the highest parts of the Popocatépetl, and that agriculture reached the lower foothills of the volcanoes (ca. 2,500–2,600 m) during the maximal developmental period of the Aztec culture in the Valley of Mexico. It is likely, however, that most of the higher areas were more or less occupied by pristine ecosystems in those periods, although pollen analysis suggests that there was some degree of deforestation before colonial times (Metclafe *et al.* 1991). Large fires could probably cause this during serious drought period.

After the conquest by the Spanish, it is assumed that land use pressure firstly decreased and that during two - three centuries the high parts of Iztac-Popo remained untouched. In the following period, most temperate forests in valleys, on flat terrain, around lakes and near large settlements were clear-cut in Central Mexico, but remained untouched in mountainous regions as the Iztac-Popo area. During the 20th century, significant changes took place, firstly by an increased, but small scale, traditional use of the natural ecosystems. Later on, because of the strong increase of the human population in the last decades of the 20th century, an intensified use of the forest communities by logging, grazing and burning has impacted the integrity of these ecosystems.

The following traditional land uses were originally practised in the Iztac-Popo area (Chavez and Trigo 1996, Rees 1996, López-Paniagua *et al.* 1996):

3.1.1   Agriculture

Arable crops (maize, beans, broad beans and pumpkins) and ornamental flowers were cultivated in small parcels in a traditional way (without artificial fertilisers, pesticides etc) on the plains and on the lower parts of the foothills of the volcanoes around the villages. Especially donkeys and horses were used for agricultural activities as ploughing.

3.1.2   Traditional forestry

The forest ecosystems were used in several ways:
-    small-scale logging of trees, especially in the neighbourhood of the villages; all material was transported by animals (donkeys & horses) (Fig. 3.6), and the cutting was done with hand saws;
-    cutting and collecting of small (dead) branches of trees and shrubs for fire wood;
-    cutting pieces out of the basal trunk of *Pinus* trees ("ocoteo"); this is done because these pieces have high resin content and thus set fire easily; however, the methods used for the extraction of the resin may

weaken the tree and increase the risks of infection by plant parasites or insects (Fig. 3.7);
- collecting of mushrooms;
- collecting of herbs and grasses.

Figure 3.6. Small-scale logging and collecting of firewood still occurs in the Iztac-Popo area (photograph by R. Bobbink).

A small part of the northwestern area was used for intensive forestry since 1948, because of a paper industry plant in San Rafael. These activities were stopped, however, in the early nineties.

3.1.3   Pastoral system

This system was characterised by grazing of small areas with cattle (for meat or milk) or with sheep. In the dry season parts of the grazing areas were burned to rejuvenate the forbs and grasses of the herb layer of the forests and other areas. This pastoral system was originally carried out directly around the villages with both low grazing pressure and low burning frequency.

Figure 3.7. Cutting small pieces out of the basal trunk of *Pinus* trees ("ocoteo") (left). This is done because these pieces with high resin content easily set fire. The weakened trees can become infected by an indigenous mistletoe (*Arcethobium spp*) (right) (photographs by R. Bobbink).

3.2   Impacts of increased land use in recent years

Although the traditional uses of the natural ecosystems of the Iztac-Popo area probably affected the species composition of some ecosystems, it is likely that the steep and higher parts of both volcanoes (> 2,700 m) were still in a near pristine state some decades ago (e.g. Rzedowski 1978; Velázquez *et al.* 2000). Moreover, none of the traditional land uses were allowed, formally, in the national park area (> 3600 m). In the last 3 – 4 decades of the 20th century the population in the urban areas and in the settlements around the Iztac-Popo volcanoes increased strongly, causing an intensification of the use of the natural ecosystems of the area. As shown in paragraph 2.2, more than 13% of the forested areas declined, or became fragmented in our 11-years research period, with a typical fragmentised pattern over the investigated area. In this section, the most likely causes of this deforestation pattern will be treated, together with the ecological consequences for their structure and functioning.

3.2.1   "Modern" forestry and pastoral systems, including burning

It is most likely that the observed deforestation of the Iztac-Popo area has been caused by the increased, partly illegal, human activities, even in the area of the National Park. Three main activities (or their combination) can be identified as the main causes of this decline in forest cover, viz.:

- strongly enhanced frequency of burning (Fig. 3.8), associated with the strong increase in livestock grazing (cattle, sheep);
- increased logging of trees, especially because of the use of modern cutting and transport means;
- increased settlements, especially at the eastern part of the area.

3.2.2   Ecological threats

The ecological functioning of the forests of the Iztac-Popo area is nowadays seriously threatened by this combination of changes in land use. The tree layer will become more open, both by the logging activities as by the (too intense and too frequent) fires. Fires seem to be a suitable tool for forest management at return intervals > 20 years, but they recur however every 1-5 year in Central Mexico (Velázquez *et al.* 2000). The reason for the frequent burning of the vegetation is to increase the palatability of the grasses in the dry season for the grazing livestock. It is expected that several tree species will decline because of selective logging, and, in addition, *Pinus* species will be favoured because of its high resistance against intense and frequent fires.

Figure 3.8. Picture of a large fire, even above the timberline, in the completely restricted area of the Iztac-Popo National Park (photograph by R. Bobbink).

Furthermore, the understory of the different forest types may be strongly impacted and the rejuvenation of several (tree) species hampered. Because of burning and associated increase in grazing activities, the understory of the forests may become dominated by annual grasses and herbs, instead of bunchgrasses, shrubs or perennial herbs (Fig. 3.9).

In addition, more open soil will develop. This may lead in the long term to a drastic decrease in typical, partly endemic plant diversity in this formerly very rich Central Mexican region (see Chapter 1). The changes in habitat quality and the increased forest fragmentation may also have severe consequences for the preservation of animal live on the slopes of these volcanoes. The effects on the populations of the endemic volcano rabbit (*Romerolagus diazi*) are well studied and significant, but discussed elsewhere in this book (Chapter 5). It is concluded that the high plant and animal diversity of the Iztac-Popo area is at high risk with continuation of the present deforestation rate of these mountainous ecosystems.

*Figure 3.9. Overvi*ew of a heavily grazed *Pinus hartwegii* stand in the Iztac-Popo area at ca.
3,500 m (photograph by R. Bobbink).

   In addition, frequent burning, the increased grazing pressure and the
logging activities will lead to soil uncovered by vegetation in the long term.
This may have severe consequences for the erodability of the slopes, and,
finally, on the important water retention capacity of the Iztac-Popo area, as
in any mountainous region (e.g. Hamilton and Bruijnzeel 1997). The risk
for erosion has been estimated for the Iztac-Popo area by a combination of
the digital terrain model of the area and a simple erosion function,
including slope, soil type, and vegetation cover. It became obvious that a
considerable part of this high volcanoes region has a high risk for erosion,
as in other mountains. However, the risk for erosion will increase when the
vegetation cover by trees or bunchgrasses declined and the soil becomes
uncovered (Morgan 1995) (Fig 3.10).

Figure 3.10. Overview of a highly eroded slopes at ca. 2800 m at the southeast part of the Iztac-Popo area (photograph by R. Bobbink).

This risk is then especially high because of the summer rain period with its high rainfall intensity that easily relocates soil particles. Consequently, the negative impacts of increased soil erosion may become astonishing in near future with the observed deforestation rates, as in many other regions in Mexico and comparable countries. It is thus clear that sustainability of the forested area is at risk at this moment, and that appropriate natural resource management is strongly needed in the Iztac-Popo area. In addition, non-fossil CO_2 emissions will strongly increase by the loss of soil organic matter and forest vegetation (see Chapter 7).

4. CONCLUDING REMARKS

As described in the previous section, deforestation of the Iztac-Popo area caused by changes in land use is a continuous threat to the ecological functioning of this mountainous area and to its characteristic biodiversity. The population in the metropolitan area of Mexico City has exponentially increased since the 1950s (Chapter 1). It is nowadays one of the most densely populated cities of the world. Furthermore, it lies in a basin (2,200 – 2,300 m) surrounded by high mountains. Because of its traffic, industries and leakage of liquefied petroleum gas, the quality of the air is badly known in the world. The smog episodes with (very) high concentrations of photochemical pollutants (especially ozone) are frequent and, especially, notorious since the 1970s till present (e.g. Miller *et al.* 1992, Blake and Rowland 1995). The impacts of this air pollution are very severe, because of the physiographic position in the valley of Mexico, surrounded by high mountains (3,000 – 4,000 m) and its climate with frequent stable atmospheric conditions. Therefore, the implications for human health are tremendous and of major importance for the government. It is thus to be expected that these (very) high ozone concentrations may affect the forests on the mountains around the town, too, because ozone is known to be one of the most deleterious pollutants for vegetation under these relatively warm, climatic conditions (e.g Ashmore 2002).

Indeed, a severe decline and mortality has been observed in many forests on the mountains in the southwest region of the Mexico City basin since the 1980s or the 1990s. The deterioration of mountainous forest stands of *Pinus hartwegii* and *Abies religiosa* is sometimes almost complete, especially in the Desierto de los Leones and the Ajusco national parks. A dendrochronological study of radial growth clearly showed that the decline in tree growth is most likely caused by this increase in ozone pollution (Alarcón et al. 1993, personal observation of the authors). Fortunately, no signs of impacts of this smog periods on the health of the trees have been observed till present in the Iztac-Popo area, probably because of the distance of the area with respect to the source of the pollution (Mexico City). However, this pollution of ozone may become a problem for the mountainous forests of the Iztac-Popo area in near future, since at several occasions the photochemical smog also reach the foot hills of the two volcanoes (Fig. 3.11). It is evident that the control of pollutant emissions in the metropolitan area is of extreme importance, for both the health of the population as the functioning of these forests ecosystems.

Figure 3.11. Photochemical smog reaching the foothills of the Iztac-Popo area. Photograph has been taken at 2950 m on the western site of the mountains (R. Bobbink).

ACKNOWLEDGEMENT

We thank Jan van Goenendael, University of Nijmegen, the Netherlands, for his financial support to the 1997 TM image.

REFERENCES

Alarcón, M. A., De Lourdes de la I.de Bauer, M., Jasso, J., Segura, G., and Zepeda, E. M. (1993). Patron de crecimiento radial en arboles de *Pinus hartwegii* afectados por contaminacion atmosferica en el suroeste del Valle de Mexico. Agrociencia serie Recursos Narurales Renovables 3, 67-80.

Almeida, L., Cleef, A. M., Herrera, A., Velázques, A., and Luna, I. (1994). El zacatonal alpino del Volcan Popocatépetl, México, y su posición en las montañas tropicales de América. Phytocoenologia 22, 391-436.

Ashmore M.R. (2002). Effects of oxidants at the whole plant and community level. In: Air pollution and plant life (2nd edition) (J.N.B. Bell and M. Threshow, eds.) pp. 89- 118. John Wiley & Sons, Chisester.

Blake, D. R. & Sherwood Rowland, F. (1995). Urban leakage of liquefied petroleum gas and its impact on Mexico City air quality. Science 269, 953-956.

Challenger, A. (1998). Utilización y conservación de los ecosistemas terrestres de México. Pasado, presente y futuro. Ibunam, México D.F.

Chávez Cortés, J. M. and Trigo Boix, N. (1996). Programa de manejo para el parque nacional Iztaccíhuatl-Popocatépetl. Universidad Autónoma Metropolitana, Unidad Xochimilco, México D.F.

Eastman, J.R. (1993). Idrisi manual. Clark University, Graduate School of Geography, Worcester, USA.

Hamilton, L. S. and Bruijnzeel, L. A. (1997). Mountain watersheds - intergrating water, soils, gravity, vegetation, and people. In: Mountains of the world - a global priority (B. Messerli and J. D. Ives, eds.) pp. 337-370. Parthenon publishing group, New York.

Hamilton, L. S., Gilmour, D. A., and Cassels, D. S. (1997). Montane forests and forestry. In: Mountains of the world - a global priority, (B. Messerli & J. D. Ives, eds.), pp. 281-311. Parthenon publishing group, New York

Harcourt , C.S. and Sayer, J.A. (1995). The conservation atlas of tropical forest: the Americas. Simon & Schuster, New York.

Hobbs, R.J. (1998). Restoration of disturbed ecosystems. In: Ecosystems of the World 16. Disturbed Ecosystems, (ed) Walker, L, pp 673-87. Elsevier, Amsterdam.

Lilesand, T.M. and Kiefer, R.W. (1994). Remote Sensing and Image Interpretation. John Wiley & Sons, Inc. 750 p.

López-Panaguia, J., Romero, F.J. and Velázques, A. (1996). Las actividades humanas y su impacto en el habitat del conejo zacatuche. In: Ecología y conservación del zatachuche (A. Velázques, F.J. Romero and J. López-Panaguia, eds.) pp. 119-132. Publicaciones scientíficas, UNAM Fondo de Cultura Económica, México DF.

Morgan, R.P.C. 1995. Soil Erosion and Conservation. Addison Wesley Longman Limited. Edinburgh, England.

Metclafe, S.E., Street-Perrot, F.A. Perrot, R.A. and Harkness, D.D. (1991). Paleolimnology of the Upper Lerma Bassin, Central Mexico: a record of climatic change and anthropogenic disturbance since 11,600 yr. BP. Journal of Paleolimnology, 5: 197-218.

Miller, P.R., Ma. De Lourdes de la I. de Bauer, M., Que vedo, N. and Hernandez, T. (1992). Comparison of ozone exposures characteristics in forested regions near Mexico City and Los Angeles. Atmospheric Environment 28, 141-148.

Rees, P. 1996. Mexico. In: Latin America and the Caribbean: a systematic and regional survey (B. W. Blouet and O. M. Blouet, eds.) pp. 207-236. Wiley and Sons, New York.

Rzedowski, J. (1978). Vegetación de México. Limusa. México D.F. , 432 pp.

Velázquez, A. (1993). Landscape Ecology of Tláloc and Pelado Volcanoes. ITC Publication No. 16, 151 pp.

Velázquez, A., Toledo, V. M., and Luna, I. (2000). Mexican temperate vegetation. In: North American terrestrial vegetation (M. G. Barbour and W. D. Billings, eds.), pp. 573-592. Cambridge University Press, Cambridge.

Chapter 4

Birds of the Iztaccíhuatl-Popocatépetl National Park and their significance for ecotourism

Alejandro Meléndez-Herrada, Nuri Trigo Boix and Aurora Chimal-Hernández
Departamento el Hombre y su Ambiente, Universidad Autónoma Metropolitana - Xochimilco, México

Key words: bird species, bird watching, ecotourism

Abstract: The interest in observing birds in their natural habitat and not confined in a cage, responds to two basic reasons; firstly, to find and appreciate a particular bird species and secondly to enjoy a walk outdoors in contact with nature. However, birth watching is more common to foreigners who visit México spending huge amounts of money and time to travel there and watch birds as their main tourist activity. The mountain range of the Valley of México is a region with an outstanding biodiversity, a situation that is particularly confirmed by the presence of birds and their environments. A severe limitation for the understanding of the diversity of birds in Mexico is the lack of inventories of local or regional birds. With this paper, it is our intention to point out the most meaningful bird species for ecotourism as well as the places that can potentially offer to the visitor of the Iztaccíhuatl and Popocatépetl volcanoes area greatest satisfactions within a bird watching activity. Ten routes along the northwestern part of the Iztaccíhuatl are recommended for bird watching. The area is selected because of it's proximity to lodging areas such as Amecameca and Tlalmanalco and also because access to the Popocatépetl is still risky due to the volcano's low but constant activity.

G.W. Heil et al. (eds.), Ecology and Man in Mexico's Central Volcanoes Area, 73–101.
© 2003 Springer Science+Business Media Dordrecht   .

1.   INTRODUCTION

In México, 1050 species of birds have been registered (Howell 1987). They are distributed in a variety of environments, ranging from the open ocean to the high mountains of Mexico. This includes deserts as well as rain forests, and among these, environments that maintain their original characteristics as well as those that have been altered by human activities. The distribution of the birds is related to the ecological behaviour of each species, which is determined by the availability of their resources and the climatic variability in each particular geographical zone.

Along the evolution of each of the bird species, the altering environment has caused morphological and physiological adaptation as a reaction to survive the environment. The behaviour adopted by each species is another factor that causes the variability of this group of animals. For this reason, we can find various sizes and shapes of the bodies, legs and peaks, as well as different colours in the feathers and soft parts - just to mention some of the more evident morphological aspects. Another manifestation of this variation are the sounds which are produced in the way of shouts, calls or songs that identify each individual as a member of a particular species.

This brief statement on the source of the distribution of bird species is the starting point in the understanding of the pleasure that people find in watching them in their natural environment. The particular characteristics of their feathers and sounds are two of the main aspects that a birdwatcher searches for.

Another characteristic that makes a bird attractive is its rarity. So, a bird which is rare, endemic (e.g. Figure 4.1) or exclusive of a certain area or habitat can be of much more interest to a birdwatcher than only for its colourful feathers or sweet sound.

The interest in observing birds in their natural habitat and not confined in a cage, responds to two basic reasons; firstly, to find and appreciate a particular bird species and secondly to enjoy a walk outdoors in contact with nature. This activity is widely distributed in Europe and North America; only in the United States of America there are about sixty million of people who constantly observe birds that they attract to their gardens and take care of. However, in México very few people practice this activity; when they do, it is mostly as an additional activity to a tourist route during some holidays. As a more formal activity, bird watching is more common to foreigners who visit México spending huge amounts of money and time to travel here and watch birds as their main tourist activity.

Figure 4.1. *Junco phaenotus* – yellow-eye junco. Endemic species, common in pine forests and grasslands (photograph by A. Meléndez).

With this in mind, we can assume that México has a great potential to develop an ecotouristic activity focused more strongly in bird watching. This can be done all over the country in places which posses the highest ornithological diversity as well as a good accessibility. Among these places, those relatively close to cities are very valuable because of the joyfulness they can offer as well as the availability of the necessary access routes by good roads. Besides these practical aspects, the visual and singing characteristics of the bird species present in a certain place and the quality and beauty of their habitats are the main criteria to be considered in selecting a natural area for bird watching.

Literature frequently highlights the wonderful biological diversity of México. It is well known that this is due to the confluence of two biogeographic zones that occur in the Transverse Neovolcanic Axis, which contain a very high diversity of flora and fauna. Due to this situation, the mountain range of the Valley of México is a region with an outstanding biodiversity, a situation that is particularly confirmed by the presence of birds and their environments. A zone that attracts the attention due to its size is the Iztaccíhuatl – Popocatépetl National Park as a representative element of the mountain system of the Valley of México, where a variety of environments give shelter to an important number of bird species (e.g. Figure 4.2).

Figure 4.2. *Cyanocitta stelleri* – steller's jay. A resident species. This bird has distinctive feathers, and likes to fly among big trees and is easy to watch (photograph by A. Meléndez).

2.   EXPLORATORY ANALYSIS

A severe limitation for the understanding of the biodiversity of Mexico is the lack of inventories of local or regional birds. It is common that the researcher has the need to elaborate his own species lists to begin with an ornithological research or that he might need to rely directly or indirectly on the available literature to estimate the bird richness, besides his personal experience. Although most researchers use mainly the direct observation and the capture of specimens, there have been no systematic efforts to cover all environments during the different seasons. Also, the time spans of the monitoring leave a great number of matters without answer, such as the notorious absence of information about reproduction and the lack of records documenting migration at a local, altitudinal and long distance levels. As a contribution to improve our ornithological knowledge, a diagnosis of the birds of the Iztaccíhuatl – Popocatépetl National Park has been elaborated which will allow having an ornithological understanding for their conservation and use in ecotourism activities.

Escalante *et al.* (1993) estimated for the Neovolcanic Axis a bird richness of 165 species, however, this number needed to be updated and more species are to be added to this list because the territory is widespread and contains an ample variety of environments (see Chapter 2). To this purpose, Meléndez and Binnqüist (1996) made a review of the literature available for the Iztaccíhuatl – Popocatépetl National Park and what they found out was that the ornithological studies are scarce, discontinuous and very restricted in time and space. These authors could gather a list of 196 bird species, which means that in the park we can find about 61% of the 320 bird species registered for the surroundings of México City (Wilson and Ceballos-Lascurrain 1993). Among the few ornithological studies done in the volcanoes, there are those from Paynter (1952), Reyes and Halffter (1975), Nocedal (1984), Gómez and Arias (1987), Lemos and Pérez (1988) and, additionally, a group of birdwatchers called "Club para la Conservación y Observación de las Aves, D. F. (1990, 1992), whose activities as a group are now stopped, also made trips in search for wild birds and observation sites.

An analysis of the species reported for the park reveals that the 197 species belong to 130 genera, 39 families and 11 orders. At the level of families, the highest number of species is in the Parulidae family with 34 species, followed by the Tyrannidae with 18, the Trochilidae with 13 and the Turdidae with 12 species. The rest of the families have from 9 to 1 species each, as can be seen in the annex I.

This taxonomic approach has certain meaning for the ornithologist but, for practical and management purposes, it is necessary to have another kind of information that can add to the understanding of the actual situation of the bird species. Information about the spatial and temporal importance of each one of the bird species is valuable for this purpose but lacking; we only have a qualitative appreciation of abundance for some of the species. In our consideration, the status of residence, the status of risk and the permitted use are aspects that may allow a better view to support ecotourism directed to the watching of birds in their natural environment.

2.1   Residence status

The widespread idea about birds being resident or migratory does not explains in a satisfactory way the residence status; however, Howell (1987) elaborated a list of the birds of México where he includes a reasonable migratory status for each species. These categories consider species that breed and are resident in a certain area, species present only or primarily in winter and species present only or primarily in summer. Another category is given to species present only or primarily as transient in spring and autumn. For all the cases indicated above, Howell distinguishes those in which the breeding population in an area is clearly smaller or more restricted than the migrant population.

In all these cases considered by Howell, endemism is an important factor so that the status of endemic or quasi-endemic species given by Navarro and Benítez (1993) is included in the characterisation of residence status.

2.2   Risk status

The degree of vulnerability of each species locates them within the most urgent conservation priorities, which is the reason why species that survive a greater risk than others makes them more likely to be selected as key conservation species. In México the Official Norm NOM-59, published in 1994 (Diario Oficial de la Federación, 1994), establishes the degree of vulnerability of the threatened and endangered Mexican species. The annex contains a list of species threatened, vulnerable, under protection and rare.

2.3   Use

The aspects stated above are very important as elements that support the conservation actions, however, the specific use of the different bird species can also influence the conservationist objectives as shown in this section.

There are two exceptional activities in the use of wild birds; the first one regards bird species that are important for hunting. This activity is regulated by the Calendario Cinegético (hunting calendar) where very few birds are susceptible of being hunted legally. In the permit Type II Doves, two species are allowed (*Zenaida asiatica* and *Z. macroura*) and in the permit Type III Other Birds, only three species (*Cyrtonyx montezumae*,

Molotrus ater and Quiscalus mexicanus). Although the birds mentioned are attractive for hunters, this activity is not allowed within the limits of the park.

The second activity to consider is the use of these animals as pets or cage birds. This activity is regulated for singing and ornamental birds in the Diario Oficial de la Federación (1994). Under this norm, many species are attractive for legal commerce of pets. Unfortunately, it is not rare to learn about illegal presence of hunters and bird catchers in the area although the impact of these activities in the bird's populations is unknown and difficult to quantify.

3.   BIRD WATCHING

The Ministry of the Environment, Natural Resources and Fisheries (SEMARNAP), recognises that the offer and demand of wildlife resources and products confers ecotourism and wildlife hiking a potentially important economic role to nature. At the same time, the Ministry emphasises that for these activities 2,402 million pesos were generated in 1997 corresponding to 5% of the total income of the tourist sector in México.

Boo (1993) stated that in order to implement an ecotouristic strategy for protected areas, the status of the natural resources has to be established, together with a cost-benefit and an offer-demand analysis of ecotourism. She also mentions the need to have inventories of wildlife places, ecosystems or species that could be tourist attractions.

The information obtained from the birds inventory mentioned above, has many of the necessary elements to support the conservation actions in the Iztaccíhuatl – Popocatépetl National Park, however, not all the information is essential to establish an ecotouristic project for the observation of birds. Wilson and Ceballos-Lascurrain (1993) mention that it is only necessary to make an ordered and qualified description of the main possible attractions for observing or appreciating bird species. These authors also suggests that the definition of the degree of ecotouristic attractiveness in any given area is subjective and will depend on the interest, the education and the preference of the visitor. The World Tourism Organisation (1992) considers that the special appeal of "star" species has to be taken into account when establishing the ecotouristic potential of an area. Other important criteria are the representative, distinctive and interesting species there.

In this sense, Heintzelman (1983) mentions that all the records and lists included in the ornithological literature must be evaluated critically if they are to be used by birdwatchers. Furthermore, personal lists of a birdwatcher often are compared with those from specialised literature. In this way, the scientific information is used for ecotourism.

Under this view, it is our intention to point out the most meaningful bird species for ecotourism as well as the places that can potentially offer visitors of the area a great pleasure within their bird watching activities. This will be done through two criteria: the attributes of the birds and those of the most accessible environments.

4.   METHODS

Based on the former, we have selected those species that could function as indicator species for the different environments in the park with respect to bird monitoring and ecotourism use. This has been formulated through two criteria: firstly, based on birds attributes and secondly by considering areas within each environment that are more accessible.

Gathering the available ornithological literature and making several intensive transects to confirm the presence of bird species, a general list was made including the most important aspects related to each species with respect to the following features: residence status, risk status, endemism and international interest, environments in which the species is found, abundance, appealing and attractiveness.

4.1   Indicator species

The objective here was to delimit the most attractive bird species from the point of view of their morphology[1] or visual attractiveness, endemism, abundance, risk category and qualitative appreciation of each species based on the feasibility of watching it in it's natural habitat (e.g. Figure 4.3). This

---

[1] The morphology of a bird is to be understood as the feathers and other structures characteristics in its various colours, tones and designs.

last aspect is obtained from the experience in the direct observation of birds in the field. To each one of these aspects, a value was assigned and those species that gathered a bigger amount of points were selected as potentially having the greatest importance for ecotourism in the area. It is likely that in other places and under other circumstances, the resulting arrangements could vary. It is important to mention that for bird watching most importance will be given to the possibility of detecting an individual by its visual appealing higher abundance and little preferred above the other characteristics.

Figure 4.3. *Basileuterus belli* – Golden-browed warbler. This is a resident species, very attractive and not shy, which inhabits the pine and broad-leaved forests (photograph by A. Meléndez).

This is particularly directed to the novice watcher since an experienced birdwatcher will be more prone to look with more emphasis to species located within a risk category such as rare and endemic; however, the birdwatcher will not despise any bird within his sight.

As Heintzelman (1983) mentions, a birdwatcher rarely considers himself an ecologist but nevertheless he gradually understands the basic principles of habitat and ecological niche. A birdwatcher will seek to gather the highest number of watched species along his life. Therefore, every new bird species watched is called "life bird" and it is annexed to the general list of the birdwatcher and in particular for his visits to a place, estate, region or country. In this way, the confirmation or personal registration of a new species is an indication of the interest of a birdwatcher.

4.2   Routes attractiveness

The second criterion is based on the highest number of species registered for each visit or tour. For the novice this means that the sites with a high probability of gathering a bigger number of species are the more attractive with respect to other similar ones which, for some reason, have less variety of species, are hard to get to, or can be risky.

To have a diagnosis of the birds of particular sites located in an area of large dimensions, such as the Iztaccíhuatl and Popocatépetl volcanoes, it requires several years of constant monitoring and is out of reach for the current ornithological knowledge. For this reason, sites selected for this chapter are those with accessibility, environmental variability and possibilities for relatively easy walking trails within the mountain topography. Therefore, the authentication of the sites is particularly important when the visitor is searching for the attractiveness with comfort and a minimum effort as the World Tourism Organisation (1992) indicates.

5.   RESULTS.

5.1   Indicator species

The species suggested as indicators for each vegetation type are listed in Table 4.1 at the end of this section. It is important to realize that there is a large number of species for some environments while for others there are only few attractive species.

From the general bird listing, at the end of this chapter, 147 species are complete residents of the area and the other 52 are migratory. From the latter, 17 species have individuals that in a bigger or lesser degree do not migrate. That is, they are partially migratory. Completely migratory or non-wintering transitory species amount to 36 (see Annex 1). The endemic species add up to 19, which represent 5% of the total of birds considered. Besides this, four species are threatened, two are under special protection and two are rare.

From this same list (Annex 1), 103 species are of international importance within the Neotropical Migratory Bird Conservation Act (USFWS, 2003). To this respect, Rappole *et al.* (1983) mention that there are 332 nearic bird species that migrate to the neotropics: 107 of which go to forested areas. This would mean that almost 40% of these nearic migrants could arrive to the Iztaccíhuatl and Popocatépetl volcanoes.

Table 4.1. Species suggested as indicators of ecotouristic interest for different vegetation types

Environment	←	Indicator bird species	→
Abies religiosa forests	Accipiter striatus		Dendrortyx macroura
	A. cooperi		Ptilogonys cinereus
Pinus hartwegii forests	Falco sparverius		Sialia mexicana
Alpine grasslands	Oriturus superciliosus		
Secondary vegetation	Columba livia		P. versicolor
	Tyrannus vociferans		Volatinia jacarina
	Mimus poliglottos		Spicella pallida
	Toxostoma curvirostre		S. atrogularis
	Cardinalis cardinalis		Chondestes gramacus
	Pipilo fuscus		Melospiza lincolnii
	Guiraca caerulea		Passer domesticus
	Passerina cyanea		
Broad leaved forests	Falco peregrinus		Salpinctes obsoletus
	Aegolius accadicus		Catherpes mexicanus
	Cinanthus latirostris		Cinclus mexicanus
	Hylocharis leucotis		Vireu griseus
	Archilochus colubris		V. belli
	Stellula calliope		Vermivora ruficapilla
	Selasphorus sasin		V. crisalis
	Empidonax fulvifrons		Icteria virens
	Sayornis nigricans		Pipilo erythrophthalmus
	Pachyramphus aglaiae		
Mixed forests	Buteo lineatus		Vermivora pinus
	Parabuteo uniccintus		V. peregrina
	Cyrtonyx montezumae		Parula americana
	Atthis eloisa		
	Selasphorus platicercus		Helmitheros vermivorus
	S. rufus		Seiurus aurocapillus
	Trogon mexicanus		S. noveborascensis
	Camptostoma imberbe		S. motacilla

Environment	←	Indicator bird species	→
	Myiopagis viridicata	Oporornis tolmiei	
	Empidonax minimus	Geothlypis nelson	
	E. hammondi	Wilsonia canadensis	
	E. oberholseri	Cardellina rubrifrons	
	Myiarchus tuberculifer	Myioborus miniatus	
	Stelgidopteryx serripenis	M. pictus	
	Hirundo pyrrhonota	Bassileuterus rufifrons	
	Aphelocoma ultamarina	B. belli	
	Parus sclateri	Piranga flava	
	Psaltriparus minimus	P. rubra	
	Campylorhynchus megalopterus	P. ludoviciana	
	C. gularis	P. bidentata	
	Polioptila caerulea	Pheucticus ludovicianus	
	Myadestes occidentalis	P. melanocephalus	
	Catharus frantzii	Alapetes pileatus	
	C. gutatus	A. virenticeps	
	Turdus assimilis	Melozone kieneri	
	T. rufopalliatus	Diglosa baritula	
	T. migratorius	Aimophila ruficeps	
	Vireo solitarius	Icterus cucullatus	
	V. gilbus	I. galbula	
	P. superciliosa	I. parisorum	
	Dendroica pensilvanica		
Induced and subalpine grassland	Eremophila alpestris	Euphagus cianocephalus	
	Sialia sialis	Molothrus aeneus	
	Lanius ludovicianus	Molothrus ater	
	Euphagus cianocephalus		
Quercus and Alnus	Zenaida aziatica	Toxostoma ocellatum	
	Z. macroura	Melanotis caerulescens	
	Geococcyx velox	Phainopepla nitens	
	Coccysus americanus	Vireo gilvus	
	Caprimulgus vociferous	Vireolanius melitophris	
	Colibri thalassinus	Vermivora celata	
	Grallaria guatimalensis	V. virginiae	
	Contopus sordidulud	Dendroica petechia	
	Empidonax traillii	D. coronata	
	Thryiomanes bewickii	D. nigrescens	
	Regulus satrapa	Wilsonia pusilla	
	R. calendula	Euphonia elegantissima	

Environment	← Indicator bird species →	
Pine forests	Catharus aurantirostris	
	Coccysus erythrophthalmus	Sitta carolinensis
	Bubo virginianus	S. pygmaea
	Glaucidium gnoma	Certhia americana
	Chordeiles minor	Catharus ustulatus
	Eugenes fulgens	Ridwayia pinicola
	Melanerpes formicivorus	D. fusca
	Sphyrapicus varius	Mniotilta varia
	Picoides scalarius	Setophaga ruticilla
	P. villosus	Ergaticus ruber
	P. stricklandi	Peucedramus taeniatus
	Colaptes auratus	Spicella passerina
	Mitrephanes phaeocercus	Carpodacus mexicanus
	Copntopus borealis	Loxia curvirostra
	C. pertinax	Carduelis pinus
	Empidonax affinis	C. notata
	E. occidentales	C. psaltria
	Tachycinet athalassina	C. tristis
	Cyanocitta stelleri	Coccothraustes abeillei
	Aphelocoma coerulescens	C. vespertinus
All environments	Coragyps atratus	Streptoprocne semicollaris
	Cathartes aura	Chaetura vauxi
	Cypseloides niger	Aeronautes saxafalis
	C. rutilus	

5.2   Important plant species for birds

Due to the different flowering and fructification stages of the plants and to the availability of many kinds of vertebrate and invertebrate animals, birds can find food practically all year round in these forests. In general, the grasses and composite species bear fruits from July to February and the species from other families have alternating flowering periods. It is interesting to note that especially in winter some species of the Labiatae family, which are particularly appealing for hummingbirds, are flowering.

Although birds can profit from the variety of seeds and flowers available at any time of the year, some especially important plants are listed below. For some of the plant species, the common name is indicated in the right column. The preference of birds to these plants is due to the exquisite nectar in the flowers or the very tasty fruits and seeds that the plants have.

Begonia gracilis	begonia
Brassica campestris	nabo
Bouvardia ternifolia	
Castilleja spp.	
Cestrum thyrsoideum	hierba del zopilote
Coriaria ruscifolia	tlalocopétatl, tlalocopetate
Eryngium carlinae	hierba del sapo
Eruca sativa	nabo, vaina para pajaros
Fragaria mexicana	fresa silvestre
Fuchsia microphylla	aretillo
Lamourouxia spp	
Lupinus ssp.	
Macroptilidium gibbosifolium	
Monnina ciliolata	
Penstemon roseus	
Phorandendron velutinum	muerdago
Prunus capuli var *serotina*	capulín
Ribes ciliatum	capulincillo
Rubus spp.	zarzamora silvestre
Salvia fulgens	
Symphoricarpos microphyllus	perlilla
Solanum cervantesii	frutilla

5.3   Proposed routes for bird watching

From an ecotouristic point of view, the following routes along the northwestern part of the Iztaccíhuatl are recommended for bird watching (Figure 4.4). The region is selected because of it's proximity to lodging areas such as Amecameca and Tlalmanalco and also because access to the Popocatépetl area is still risky due to the volcano's low but constant activity.

The routes are traced in the map and numbers here refer to those of the map. The vegetation types present are listed according to the characterisation described in Chapter 3.

Route 1. Amecameca - Paso de Cortés - Tlamacas.

Leaving from Amecameca (A) this route is the most easy to drive because it has a paved road, which is constantly being maintained. Along the road, once you get to the 2800 masl, pine forests dominate. Later on, where the road goes along a ravine, fir dominates. In the higher part, near the 3200 masl, *Pinus hartwegii* forests begin and they continue to dominate up to the 3700 masl at Paso de Cortés where the alpine grassland is located. If the volcano ceases it's activity, it is possible to reach higher to the alpine facilities in Tlamacas.

Route 2. Zoquiapan.

This route is reached from the México-Puebla highway (11 in the map) and is usually drivable as far as the Zoquiapan Station that is managed by the Universidad Autónoma de Chapingo. From there, depending on the season, you might be able to drive further south for 2-4 more kilometres but this hill-lands area is very rewarding for an easy hike. A variety of environments can be easily reached along this route: mixed forests, pure oak and pure fir areas as well as some spot with *Pinus hartwegii* and alpine grasslands.

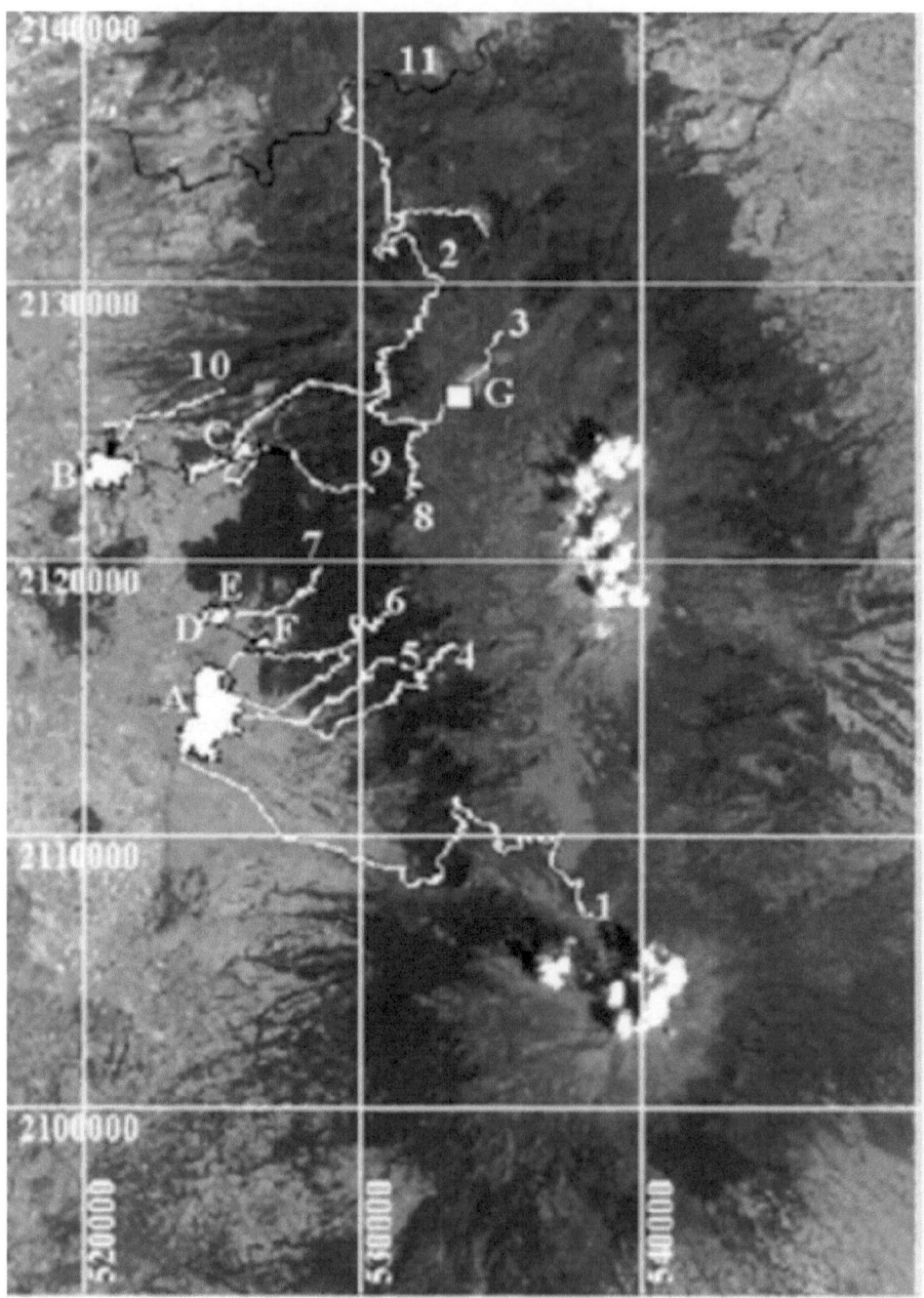

Figure 4.4. Map with birds routes

Route 3. San Rafael.

Due to recent forest exploitation in the municipality of Tlalmanalco, the road that leads to this area is being improved and quite well conserved; it departs from the Town of San Rafael (C) which is reached from Tlalmanalco (B). On one hand this means more people around but possibilities for going into some relatively isolated mixed and fir forests along the 3000 to 3400 masl range are great if you take one of the many trails which begin along the road either heading north or south. If you go higher you will find alpine and sub alpine grasslands well represented as well as *Pinus hartwegii* forests. This road also allows you to go through the Llano Grande (G) towards the east of the Iztaccíhuatl to some pleasing and silent mixed and broad-leaved forests on the side of Puebla.

Route 4. Southeast from Amecameca.

Leaving from downtown Amecameca (A) to the east on a not so good road, which is driveable only during the dry season, several interesting communities host many beautiful bird species all year round. On the lower parts, a very interesting secondary shrub community is found. As you go up, mixed forest dominate with pines and broad-leaved species and in the highest part, *Pinus hartwegii* forests mingle with induced and sub alpine grasslands.

Route 5. Northeast from Amecameca.

This route is similar to the former one in it's lower part but besides the mixed forests at 3000 masl, you can also see very well preserved areas of fir forest. The road is quite drivable in the dry season and although it is not so safe for a vehicle in the rainy season, it is worth a walk.

Route 6. Tzoquitzinco and El Salto.

This route has many possibilities and is best if taken as a circuit beginning in the town of Santiago Cuautenco (F) and going upwards to see some nice mixed forests. If you are a good hiker, you can go further high until you arrive to some nice *P. hartwegii* area around the 3600 masl.

Afterwards, going back on the same trail you can take an alternative trail back by going south-west at the 2860 masl and crossing the cañada Tzoquinzinco to arrive to El Salto from where you can easily take the road back to Amecameca.

Route 7. Tlapacoya and Chopanac.

The condition of this road is better than the latter but even in the dry season you have to prepare yourself for walking at the highest part (beyond 3200 masl). The most rare community, however, which is the mountain cloud forest, is reached easily by driving. Some nice areas of mixed forest are also at walking distance near the road. This route can be accesed through Santa Isabel Chalma (E) which can be reached either from Santiago Cuautenco (F) or from San Antonio Tlaltecahuacan (D).

Route 8. High Nexcoalanco.

Following Route 3, a small road turns southwards just befor arriving to Llano Grande (G) at 3200 masl. From there it can be worth having an easy walk around the ravine to enjoy the birds feeding from the herb and shrubs species in the fir forests. If you are interested in going higher, the ravine starts at some quite well preserved *Pinus hartwegii* forests.

Route 9. Agua dulce – El Negro.

This route is exclusively for walking and can be a paradise for hikers. The trail goes on the low part of Nexcoalanco, which is one of the longest and deepest ravines in the area. Because of its quietness and the high walls of the ravine, you can enjoy a symphony of bird songs.

Route 10. Cañada del Agua.

Although this route does not go very high, it is a very attractive one from the perspective of bird watching and it is among the most accessible ones. Beginning at a road north of Tlalmanalco (B), the trail goes along a ravine called Cañada del Agua, which comes down from the north-western part of the Iztaccíhuatl. The oak and mixed forests are a wonderful home for many beautiful bird species characteristic to this type of vegetation (e.g.Figure 4.5).

Figure 4.5. *Pipilo erythrophthalmus*. Small bird that likes to fly in broad leaved forests (photograph by A. Meléndez).

6. CONCLUSIONS

This Chapter makes notice of the possibility that, through the interest for birds and bird watching, the mountain environments of the Iztaccíhuatl and Popocatépetl volcanoes can be conserved. Ornithological research and the pleasure of bird watching as an ecotouristic activity are necessary as important steps towards the establishment of knowledge and of positive attitudes towards conservation and sound use of this kind of fauna. The application of monitoring techniques strategically designed could contribute to further decision making for the use of natural resources. This has to be done within a short time as is also mentioned in Chapter 8.

Recreational and educational activities are highly recommended as a non-destructive alternative that can yield valuable economic benefits. To make use of the birds as an appeal towards this end has proven its usefulness in North America, in Europe and even in some regions of Latin America where ornithological attractiveness is the main interest for tourism, environmental education and research. The Iztaccíhuatl and Popocatépetl volcanoes offer a great variety of birds in different environments for at least four adjacent estates and, of course, for the biggest city on Earth.

There is much left to do in the area; the ornithological research has focused mainly to the West and Northwest part of the Iztaccíhuatl, without a continuous monitoring and also neglecting the aspects of habitat use, reproduction and migration. With respect to recreation, the accessibility of the existing paved road towards Tlamacas has made the Paso de Cortés area the most visited, however, other routes which are even more attractive are little known although their unpaved roads can be easily driven during the dry season.

Finally, there is still much research to be done about the effects that the current activity of the Popocatépetl will bring about as well as the effects of the heavy drought and fires registered during 1998 as an effect of the "El Niño" and the more permanent and long lasting legal and illegal logging in the area.

REFERENCES

American Ornithologists Union (1998) Checklist of North American Birds. http://www.aou.org/aou/birdlist.html Access date: April 26, 2003.

Boo, E. (1993) Ecotourism planning for protected areas.15-31p. In: K. Lindemberg and D. E. Hawkings (Eds.). Ecotourism: a guide for planners & managers. The Ecotourism Society, North Bennington, Vermont.

Brandon, K. (1993). Basic steps toward encouraging local participation in nature tourism projects. 134-151p. In: K. Lindemberg and D. E. Hawkings (Eds.). Ecotourism: a guide for planners & managers. The Ecotourism Society, North Bennington, Vermont.

Club para la Conservación y Observación de las Aves, D. F. (1990). Aves de la zona del Parque Izta-Popo: abarca desde Popo Park hasta Tlamacas. Notas sin publicar del Club (CCOADF). México.

Club para la Conservación y Observación de las Aves, D. F. (1992). Registros: del 9/XI/91 al 1/II/92. Boletín del CCOADF (11): 2. México.

Diario Oficial de la Federación. (1994). Norma Oficial Mexicana NOM-059-ECOL-1994, que determina las especies y subespecies de flora y fauna silvestres terrestres y acuáticas en peligro de extinción, amenazadas, raras y las sujetas a protección especial y que establece especificaciones para su protección. Diario Oficial de la Federación, Organo del Gobierno Constitucional de los Estados Unidos Mexicanos. Tomo CDLXXXVIII, No. 10, México, D.F. lunes 16 de mayo de 1994.

Escalante Pliego, P., Navarro Sigüenza, A. G. and Peterson, A. T. (1993). A geographic, ecological, and historical analysis of land bird diversity in Mexico. p281-289. In: T. P. Ramamoorthy, R. Bye, A. Lot and J. Fa. Biological diversity of Mexico: origins and distribution. Oxford University Press, New York.

Furness, R. W. and Greenwood, J. J. D. (Editors) (1993). Birds as monitors of environmental change. Chapman & Hall, London 356pp.

Gómez, G. and Arias, P.. (1987). Estudio de la avifauna de los volcanes La Malinche y Popocatépetl. p.212-225. En: Memorias del V Simposio de Fauna Silvestre, México.

Heintzelman, D. S. (1983). The birdwatcher's activity book. Stackpole Books, Harrisburg PA 54p.

Howell, S.N.G. (1987). A field checklist to the birds of Mexico. Golden Gate Audubon Society. Berkeley, California 21pp.

Lemos Espinal, J. A. and Pérez Monroy, A. (1988). Estructuración de una comunidad de aves en un pastizal de la vertiente oriental del volcán Iztaccíhuatl, Puebla. p.245-262. In: Memorias del Segundo Simposio Internacional de Vida Silvestre, Acapulco, Gro. Wildlife Society – SEDUE, México.

Meléndez Herrada, A. and Binnqüist Cervantes, G..(1996). Diagnóstico de la avifauna. 91-96p. In:. Programa de Manejo para el Parque Nacional Iztaccíhuatl-Popocatépetl (Chávez Cortés, J.M. and Trigo Boix, N., Coords.). Colección Ecología y Planeación, Universidad Autónoma Metropolitana, Unidad Xochimilco.

Navarro, A.G. and Benítez, H. (1993). Patrones de riqueza y endemismo de las aves. Ciencias, número especial 7:45-54. México.

Nocedal, J. (1984). Estructura y utilización del follaje por las comunidades de pájaros en bosques templados del Valle de México. Acta Zoológica Mexicana (6): 1-45.

Partners in Flight Program (1992). Preliminary list of migrants for Partners in Flight neotropical migratory birds conservation program. Infrormation and Education Working Group and the National Fish and Wildlife Foundation, Annual Report 1991, 2(1): 30

Paynter, R. A., Jr. (1952). Birds from Popocatepetl and Iztaccihuatl, Mexico. Auk 70(3): 338-349.

Rappole, J. H., Morton, E. S., Lovejoy, T. E. and Ruos, J. L. (1983). Nearctic avian migrants in the neotropics. U. S. Department of the Interior, Fish and wildlife Service, Washington 646p.

Reyes Castillo, P. and Halfter, G. (1975). Fauna de la cuenca del valle de México. p.135-180. In: Departamento del Distrito Federal. Memorias de las Obras del Sistema de Drenaje Profundo del Distrito Federal, Tomo 1. México.

USFWS (2003). Neotropical Migratory Bird Conservation Act. Division of Bird Habitat Conservation. United States Fish and Wildlife Service. http://birdhabitat.fws.gov/NMBCA/eng_birdlist.htm Date of last access, April 26, 2003.

Wilson, R.G. and Ceballos-Lascurain, H.. (1993). The birds of Mexico City. BBC Printing and Graphics Ltd. 2a. ed.

World Tourism Organization (1992). Guidelines: development of national parks and protected areas for tourism. WTO/UNEP Technical Report Series No. 13 Madrid 53p.

APPENDIX I

List of the Birds in the Iztaccíhuatl – Popocatépetl volcanoes Birds are listed according to the taxonomic order of the American Ornithologists' Union (1983), the residence status was taken from Howell (1987); the risk status was taken from the Diario Oficial de la Federación (1994); endemism or quasi-endemism from Navarro y Benítez (1994); and the international interest from the Neotropical Migratory bird Conservation Act (USFWS 2003).

Keys
1. Residence status taken from Howell (1987):
*   Breeds (used alone denotes a species resident within México, although breeding and no breeding ranges may differ; used before a different status – W, S or T1- indicates that breeding population is clearly smaller/more restricted than migrant population.)
W   Present only/primarily in winter.
S   Present only/primarily in summer
T1   Present only/primarily as a transient in spring and/or autumn

2. Risk status, endemism and international interest:
Th   Threatened
Pr   Special protection
R   Rare
E   Endemic or quasi endemic
I   Of international interest as neotropical migratory bird

3. Environments:
Ar   Abies religiosa
Ph   Pinus hartwegii
Ag   Alpine grassland
Sv   Secondary vegetation
Mf   Mixed forest
Is   Induced and Sub alpine grassland
Bl   Broad leaves
QA   Quercus spp. and Alnus spp.
Pi   Pines
Q   Quercus spp
Ae   All environments

4. Abundance:
C   Common
FC   Fairly Common
UC   Uncommon
R   Rare
VR   Very Rare

5. Appealing:
+++   Very appealing
++   Appealing

+   Little appealing
-    Non appealing, cryptic or inconspicuous

6. Attractiveness:
+++   Very attractive
++   Attractive
+   Little attractive
-     Not attractive

Taxon   Keys

	1	2	3	4	5	6
CICONIIFORMES						
Cathartidae						
Coragyps atratus	*	I	Ae	VR	-	-
Cathartes aura	*	I	Ae	VR	-	-
FALCONIFORMES						
Accipitridae						
Accipiter striatus	*	Th, I	Ar, Sv	R	++	++
Accipiter cooperii	*	I	Ar, Sv	R	++	++
Parabuteo unicinctus	*	Th, I	Mf, Bl, QA	R	++	++
Buteo lineatus	*W	-	Mf, Bl, QA	VR	++	+
Buteo jamaicensis	*	Pr, I	TA	C	++	+++
Falconidae						
Falco sparverius	*W	I	Ph, Is, Sv	C	++	++
Falco peregrinus	*W	Th ,I	Bl, Pi, Is	VR	++	+++
GALLIFORMES						
Odontophoridae						
Dendrortyx macroura	*	Pr, E	Ar, Bl	R	-	+++
Cyrtonyx montezumae	*	-	Mf, Bl	VR	-	+
COLUMBIFORMES						
Columbidae						
Columba livia	*	-	Sv, Is	C	+	-
Zenaida asiatica	*	I	QA, Sv	R	++	++
Zenaida macroura	*	I	QA, Sv	R	++	++
Columbina inca	*	-	Sv	C	+	+

CUCULIFORMES

Cuculidae

Coccyzus erythrophthalmus	T1	I	Pi, QA, Sv	R	++	+
Coccyzus americanus	*S	I	QA, Pi, Sv	R	++	+
Geococcyx velox	*	-	QA, Sv	VR	++	+

STRIGIFORMES

Strigidae

Bubo virginianus	*	Th	Pi, Ph, QA	R	-	++
Glaucidium gnoma	*	R	Pi, Ph, QA	R	-	++
Aegolius acadicus	*	-	Bl, Pi	R	-	+

CAPRIMULGIFORMES

Caprimulgidae

Chordeiles minor	*S	I	Pi, Bl	R	-	+
Caprimulgus vociferus	*	I	QA, Bl	R	-	++

APODIFORMES

Apodidae

Cypseloides niger	*	I	Ae	R	-	+
Cypseloides rutilus	*	-	Ae	R	-	+
Streptoprocne rutila	*	-	Ae	R	-	+
Streptoprocne semicollaris	*	E	Ae	R	-	+
Chaetura vauxi	*	I	Ae	R	-	+
Aeronautes saxatalis	*	I	Ae	R	-	+

Trochilidae

Colibri thalassinus	*	-	QA, Bl	FC	++	++
Cynanthus latirostris	*	I	Bl, Sv	C	++	++
Hylocharis leucotis	*	-	Bl, QA, Sv	C	++	++
Amazilia beryllina	*	-	Sv, QA	FC	++	++
Lampornis clemenciae	*	I	Sv, Bl	FC	++	++
Eugenes fulgens	*	I	Pi, Sv	FC	++	++
Calothorax lucifer	*	-	Sv	R	++	++
Archilochus colubris	W	I	Bl, QA, Sv	FC	++	++
Stellula calliope	*W	I	Bl, QA, Sv	R	++	++
Atthis heloisa	*	E	Mf, QA	VR	++	+
Selasphorus platycercus	*	I	Mf	FC	++	++
Selasphorus rufus	W	I	Mf	FC	++	++

Selasphorus sasin	W	I	Bl, Ph	VR	++	+

TROGONIFORMES

Trogonidae

Trogon mexicanus	*	-	Mf, QA	R	+++	++

PICIFORMES

Picidae

Melanerpes formicivorus	*	-	Pi, QA,	C	++	+++
Sphyrapicus varius	W	I	Pi, QA	R	++	++
Picoides scalarius	*	-	Pi, QA	C	++	++
Picoides villosus	*	-	Pi, QA	FC	++	++
Picoides stricklandi	*	E	Pi, Ph, QA	C	+	+++
Colaptes auratus	*	-	Pi, Ph	C	+++	+++

PASSERIFORMES

Formicariidae

Grallaria guatimalensis	*	-	QA, Ar	VR	-	+

Tyrannidae

Camptostoma imberbe	*	I	Mf, Sv	R	-	+
Myiopagis viridicata	*	-	Mf, Sv	R	-	+
Mitrephanes phaeocercus	*	-	Pi, Mf	UC	++	++
Contopus cooperi	*T1	-	Pi, Mf	R	-	+
Contopus pertinax	*	I	Pi, Sv	C	-	+
Contopus sordidulus	*S	I	QA, Pi	FC	-	+
Empidonax traillii	*W	I	QA, Sv	VR	-	+
Empidonax minimus	W	I	Mf, Bl	R	-	+
Empidonax hammondii	W	I	Mf, Bl	FC	-	+
Empidonax oberholseri	W	I	Mf, Pi	FC	-	+
Empidonax affinis	*	-	Pi	R	-	+
Empidonax occidentalis	*	I	Pi, QA	C	-	+
Empidonax fulvifrons	*	I	Bl, Mf	C	++	++
Sayornis nigricans	*	-	Bl	VR	++	++
Pyrocephalus rubinus	*	I	Is, Sv	C	+++	+++
Myiarchus tuberculifer	*	I	Mf, Sv	VR	+	+
Tyrannus vociferans	*	I	Sv, Is	R	+	+
Pachyramphus aglaiae	*	I	Bl, Mf	VR	+++	++

Laniidae

Lanius ludovicianus	*	I	Is, Sv	C	++	++

Vireonidae

Vireo griseus	*	I	Bl, Sv	VR	+	+
Vireo bellii	*	I	Bl, Sv	R	+	++
Vireo solitarius	*	I	Mf, QA	C	+	+
Vireo huttoni	*	-	Q, Mf,	C	+	+
Vireo gilvus	*	I	Mf	C	+	+
Vireolanius melitophrys	*	-	Q	VR	++	+

Corvidae

Cyanocitta stelleri	*	-	Pi, Mf	C	+++	+++
Aphelocoma coerulescens	*	-	Pi, Mf	FC	+++	+++
Aphelocoma ultramarina	*	-	Mf, Bl	R	++	+
Corvus corax	*	-	Ph	R	++	++

Alaudidae

Eremophila alpestris	*	-	Is	R	++	++

Hirundinidae

Tachycineta thalassina	*	I	Pi, Mf	C	++	++
Stelgidopteryx serripenis	*	I	Mf, Sv	R	+	+
Petrochelidon pyrrhonota	*S	I	Mf, Bl	R	++	+
Hirundo rustica	*	I	Sv, Mf	C	+++	++

Paridae

Poecile sclateri	*	E	Mf, Pi,	UC	++	++

Aegithalidae

Psaltriparus minimus	*	-	Mf, Bl, QA	C	+	++

Sittidae

Sitta carolinensis	*	-	Pi, Mf	R	++	+
Sitta pygmaea	*	-	Pi, Ph	FC	++	++

Certhidae

Certhia americana	*	-	Pi	FC	++	++

Troglodytidae

Campylorhynchus megalopterus	*	E	Mf, QA	FC	+++	+++
Campylorhynchus gularis	*	E	Mf	VR	+	+
Salpinctes obsoletus	*	-	Bl	R	++	+
Catherpes mexicanus	*	-	Bl	FC	++	++
Thryomanes bewickii	*	-	QA, Sv	C	++	++
Troglodytes aedon	*	I	Pi, Ar, Mf	C	+	++
Cistothorus platensis	*	I	Ph, Is	VR	+	+

Cinclidae

Ciclus mexicanus	*	-	Bl	VR	+	++

Regulidae

Regulus satrapa	*	-	QA, Bl, Pi	C	++	+++
Regulus calendula	*W	I	QA, Bl, Pi	C	+	++

Silviidae

Polioptila caerulea	*	I	Mf, Q, Bl	C	++	++

Turdidae

Sialia sialis	*	I	Is, Ph, Ag	R	++	++
Sialia mexicana	*	I	Ph, Ag, Is	C	++	++
Myadestes occidentalis	*	-	Mf, Ar	C	++	+++
Catharus aurantirostris	*	-	QA, Bl, Ar	C	+	+
Catharus occidentalis	*	E	Mf, QA	C	+	++
Catharus frantzii	*	-	Mf, Bl	R	+	+
Catharus ustulatus	W	I	Pi, Ar	VR	+	+
Catharus guttatus	W	-	Mf, Pi	FC	+	++
Turdus assimilis	*	-	Mf, Bl	R	++	++
Turdus rufopalliatus	*	E	Mf, QA	C	++	+++
Turdus migratorius	*	I	Mf, Ar, Bl	C	+++	+++
Ridgwayia pinicola	*	-	Pi, Q	VR	++	+

Mimidae

Mimus polyglottos	*	-	Sv, Is	R	++	++
Toxostoma ocellatum	*	E	Q, Mf	R	+	++
Toxostoma curvirostre	*	-	Sv, Bl	C	++	++
Melanotis caerulescens	*	E	QA, Sv	C	++	+++

Bombycillidae

Bombycilla cedrorum	W	-	Pi, Ar	R	++	++

Ptilogonatidae

Ptilogonys cinereus	*	-	Ar, Mf	C	++	++
Phainopepla nitens	*	-	Q	VR	+	+

Peucedramidae

Peucedramus taeniatus	*	I	Pi, Mf	C	+	++

Parulidae

Vermivora pinus	W	I	Mf, Sv	VR	++	++
Vermivora peregrina	W	I	Mf, Sv	VR	++	++
Vermivora celata	W	I	QA, Mf	C	++	++
Vermivora ruficapilla	W	I	Bl, Sv	C	+	+
Vermivora virginiae	W	I	Q, Bl	R	++	++
Vermivora crissalis	*	I	Bl, QA	R	++	++
Parula superciliosa	*	-	Mf	C	+++	++
Parula americana	W	I	Mf	R	++	++
Dendroica petechia	*	I	QA, Sv	R	++	++

Dendroica pensylvanica	T1	I	Mf	VR	++	+
Dendroica magnolia	W	I	Pi	VR	++	+
Dendroica coronata	*W	I	QA, Bl, Pi	C	++	++
Dendroica nigrescens	*W	I	Q, Mf	C	++	++
Dendroica virens	W	R, I	Pi, Mf	R	++	+++
Dendroica townsendi	W	I	Pi, Mf	C	++	+++
Dendroica occidentalis	W	I	Pi, Mf	C	++	+++
Dendroica fusca	T1	I	Pi, Mf	VR	++	+
Mniotilta varia	W	I	Pi, Mf	C	++	+++
Setophaga ruticilla	W	I	Pi, Mf	R	++	++
Helmitheros vermivorus	W	I	Mf QA	VR	+	+
Seiurus aurocapillus	W	I	Mf, QA	R	+	+
Seiurus noveborascensis	W	I	Mf, QA	R	+	+
Seiurus motacilla	W	I	Mf, QA	R	+	+
Oporornis tolmiei	*W	I	Mf, QA	C	++	++
Geothlypis nelsoni	*	E	Mf, QA, Bl	C	++	+++
Wilsonia pusilla	W	I	QA, Mf, Sv	C	++	++
Wilsonia canadensis	T1	I	Mf	R	++	+
Cardellina rubrifrons	*	I	Mf, QA	R	++	++
Ergaticus ruber	*	E	Pi, Mf	C	+++	+++
Myioborus pictus	*	I	Mf, Pi, Q	R	+++	++
Myioborus miniatus	*	-	Mf, Pi, Q	C	+++	+++
Basileuterus rufifrons	*	-	Mf, Sv	C	+	++
Basileuterus belli	*	-	Mf	C	+	++
Icteria virens	*	-	Bl	VR	++	+

Thraupidae

Piranga flava	*	I	Mf, QA	C	+++	++
Piranga rubra	*	I	Mf, Sv	C	+++	++
Piranga ludoviciana	*W	I	Mf, QA, Bl	R	++	++
Piranga bidentata	*	-	Mf	VR	++	+
Euphonia elegantissima	*	-	Q, Bl	R	+++	++

Emberizidae

Volatinia jacarina	*	-	Sv	VR	+	+
Diglossa baritula	*	-	Mf, Sv	UC	++	++
Atlapetes pileatus	*	E	Mf, Q	C	++	+++
Buarremon virenticeps	*	E	Mf, QA	C	++	+++
Melozone kieneri	*	E	Mf, Bl	R	++	++
Pipilo erythrophthalmus	*	-	Bl, QA, Pi	C	++	++
Pipilo fuscus	*	-	Sv, Mf	C	+	+
Aimophila ruficeps	*	-	Mf, QA	R	++	+
Oriturus superciliosus	*	E	Ag, Ph	C	++	+++
Spizella passerina	*	I	Pi	C	++	++
Spizella pallida	W	I	Sv, Is	R	+	+
Spizella atrogularis	*	I	Sv, Is	R	+	+
Chondestes grammacus	*W	I	Sv, Is	C	+	+
Xenospiza baileyi	*	E	Ag, Ph	VR	+	++

| Melospiza lincolnii | W | I | Sv, Is | C | + | + |
| Junco phaeonotus | * | E | Pi, Ph, Ag | C | ++ | ++ |

Cardinalidae

Cardinalis cardinalis	*	-	Sv	VR	+++	+
Pheucticus ludovicianus	W	I	Mf, QA	VR	++	++
Pheucticus melanocephalus	*	I	Mf, QA	C	++	+++
Guiraca caerulea	*	I	Sv, Is	C	++	++
Passerina cyanea	W	I	Sv, Is	R	++	++
Passerina versicolor	*	I	Sv, Is	R	++	++

Icteridae

Euphagus cianocephalus	W	-	Is, Sv	R	+	-
Quiscalus mexicanus	*	-	Is, Sv	C	+	-
Molothrus aeneus	*	I	Is, Sv	C	+	-
Molothrus ater	*	I	Is, Sv	C	+	-
Icterus cucullatus	*	I	Mf, QA	R	++	+
Icterus galbula	*	I	Mf, QA	C	++	++
Icterus parisorum	*	I	Mf, Q	R	++	++

Fringillidae

Carpodacus mexicanus	*	-	Pi, Sv, Mf	C	++	++
Loxia curvirostra	*	-	Pi	R	++	++
Carduelis pinus	*	-	Pi, Ph, Ag	R	+	++
Carduelis notata	*	-	Pi, Sv	R	++	+
Carduelis psaltria	*	I	Pi, Sv	C	++	++
Carduelis tristis	W	I	Pi, Sv	VR	++	+
Coccothraustes abeillei	*	-	Pi	R	++	+
Coccotraustes vespertinus	*	-	Pi	R	++	+

Passeridae

| Passer domesticus | * | - | Sv | C | + | - |

Chapter 5

Effects of habitat fragmentation on the mammalian assemblage at the Iztaccíhuatl and Popocatépetl Volcanoes, Mexico

Alejandro Velázquez*, Francisco J. Romero**, Héctor Rangel-Cordero* and Gerrit W. Heil***

*Instituto de Geografia, UNAM, México **Laboratoria de Ecología y Conservacion de Fauna Silvestre, Departamento El Hombre y su Ambiente, UAM-Xochimilco, ***Department of Plant Ecology,Faculty of Biology, Utrecht University, The Netherlands

Key words: conservation, landscape changes, mammals, RS/GIS, volcanoes

Abstract: Rapid habitat transformation calls for efficient methods to lead conservation efforts. For this reason, landscape analysis is becoming a major issue in biodiversity conservation. In Mexico, as in many parts of the world, management strategies are scarce so that biodiversity depletion processes continue. This is the case on the Iztaccíhuatl and Popocatépetl volcanoes, which harbour over 10% of the total Mexican mammalian species within ca. 0.02% of the surface of Mexico. The present paper aims at assessing the effect of landscape dynamics on mammalian assemblages through a RS/GIS modelling approach. A database including all mammalian species recorded in the region from 1839 up to 1997 was compiled. The records, at genus level, were linked to land cover classes obtained from Landsat satellite images taken in 1986 and 1997. Land cover and habitat changes were analysed through a statistical analysis by crossing land cover maps of 1986 and 1997, which were transformed, into habitat richness types. Major changes from high to medium and medium to low habitat richness classes prevail in the area. This reduces the resilience of the natural landscapes and increases the threats for most mammalian species. From a landscape ecological perspective, the present chapter demonstrates the importance of the area as a unique mosaic of mammalian assemblages.

G.W. Heil et al. (eds.), Ecology and Man in Mexico's Central Volcanoes Area, 103–123.
© 2003 Springer Science+Business Media Dordrecht

1.   INTRODUCTION

The current rate of habitat transformation and depletion calls for efficient methods to guide conservation efforts (Ehrlich and Wilson 1991, Myers 1993). To achieve this, large databases are usually needed to monitor influences on landscape changes and implicit natural habitat declining (Hoth *et al.* 1985; Schwartz 1993). Nonetheless, the simultaneous study of plant–animal assemblages and the effect on their distribution patterns has rarely been considered for understanding landscape dynamics (Velázquez and Heil 1996; Velázquez *et al.* 2001). Nowadays, the understanding of driving processes at landscape level is becoming a major issue in land use planning and biodiversity conservation (Shilling 1997; Velázquez and Romero 1999). This implies the use of spatially explicit models on which to build sound restoration, conservation and management actions. This is most relevant within protected areas, such as the Iztaccíhuatl and Popocatépetl volcanoes, which harbor much biodiversity and endemic taxa (Myers 1993; Chávez and Trigo 1996; Maza and Soberón 1998). In Mexico, for instance, most protected areas are experiencing rapid habitat changes (Dirzo and Gomez-Pompa 1997). In these areas, management strategies are scarce so that biodiversity depletion processes still dominate (Mittermeier and Mittermeier 1998). This is the case on the Iztaccíhuatl and Popocatépetl volcanoes, which harbor over 10% of the total Mexican mammalian species within ca. 0.02% of the national surface area. The aim of this Chapter is twofold. First, to analyze landscape changes over a period of 10 years (1986–1997), and second to evaluate how these changes affect mammalian assemblages. It is expected that small- and medium-sized mammals are more sensitive to landscape changes than large body sized mammals *(sensu* Brown 1995). Additional considerations are given in the light of conservation and management of the study area.

2.   METHOD

2.1   Study area

The Iztaccíhuatl and Popocatépetl Volcanoes (18°59' and 19°04 N and 98°42 and 98°30 W) form part of the Transverse Neovolcanic Axis and are the core areas of the Izta-Popo National Park (DETENAL 1983). They are located about 45 km east of Mexico City (Figure 5.1). Geologically, these strato-volcanoes have been formed in sequential stages of volcanic activity, starting in the Miocene with a maximum development in the Pleistocene (Demant 1978; Lugo 1984). Both volcanoes have a pronounced altitudinal gradient, ranging from 2,300 to 5,452 m.

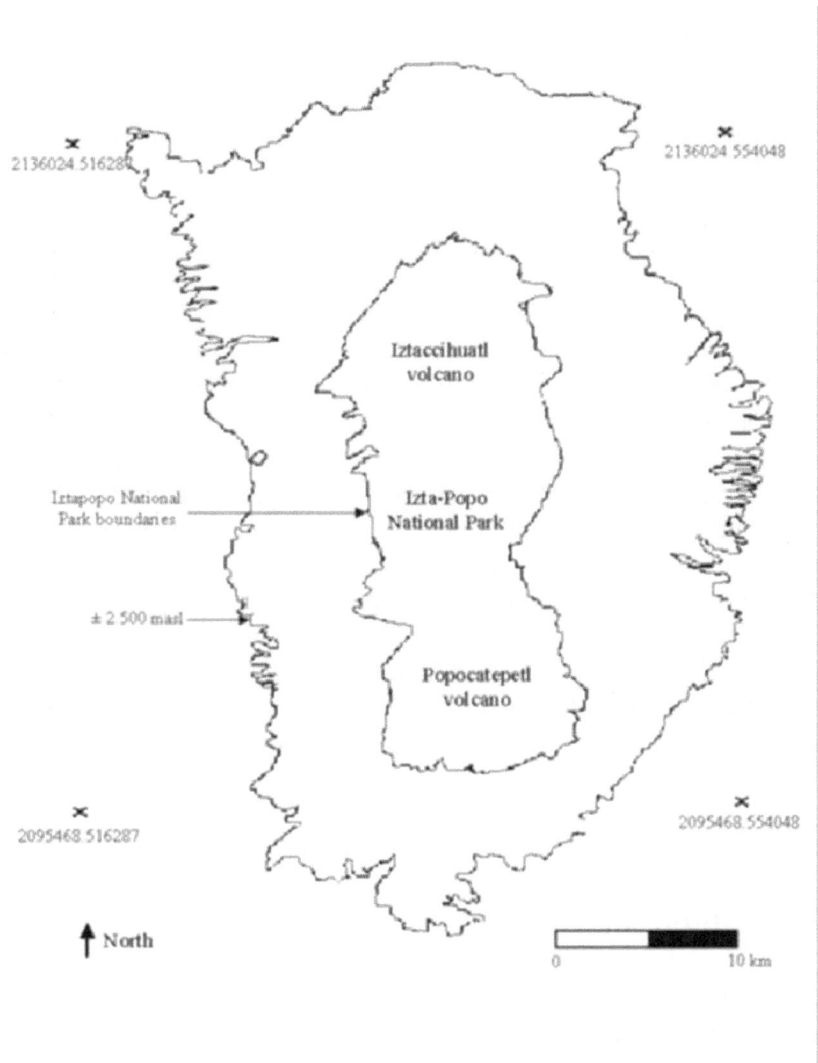

Figure 5.1. Location of the study area. The Iztaccíhuatl and Popocatépetl volcanoes are found in central Mexico southeast of Mexico City.

The climate varies from temperate at 2625 m, to cold at 4000 m types (García 1981). Andosol, Regosol and Lithosol soil types prevail. Temperate coniferous forest dominates between 2,850 and 3,850 m and in its upper limits there is contact with alpine bunch grassland at ca. 4,000 m (Almeida et al. 1994; Chávez and Trigo 1996).

2.2   Compilation of mammalian records

A database was compiled including all mammalian species recorded in the region from 1839 up to 1997. Hamilton-Smith (1839), Saussure (1860), Ferrari-Pérez (1886) and Merriam (1895, 1898) conducted the most systematic surveys during the nineteenth century. Nelson and Goldman (1934), Rojas (1951) and Villa-Ramírez (1953) provided a complete list of the mammals of the region in the last century.

This compilation was needed to understand historical distribution patterns of some taxa, as well as to include data from collections (Instituto de Biología, UNAM and Facultad de Ciencias Biológicas, IPN) and bibliographic references (Santillán 1979; Blanco *et al.* 1981; Ceballos and Galindo 1984; Hoth *et al.* 1987). In addition, field data taken by the authors from 1985 until 1997 were also incorporated in the final database. These data include trapping (over 25 days in periods of three consecutive nights using 200 Sherman traps), indirect observations (e.g., foot-prints, carcasses, pellets) and interviews with local inhabitants (hunters and shepherds), as suggested by Sutherland (1996). All information was then pooled into a final database which included all geo-referenced locations of records as accurately as possible so that links between species records and land coverage classes was likely to be conducted *(sensu* Velázquez 1993; Figure 5.2).

2.3   Habitat analysis

Two methods were used to characterize habitat attributes *(sensu* Velázquez 1993). The first followed field sampling of vascular plant species per major vegetation type. These vegetation types were depicted from phyto-sociologic surveys that included sampling of 185 relevés (Mueller-Dombois and Ellenberg 1974).

Figure 5.2. Flow chart showing the process followed to achieve the final results described in

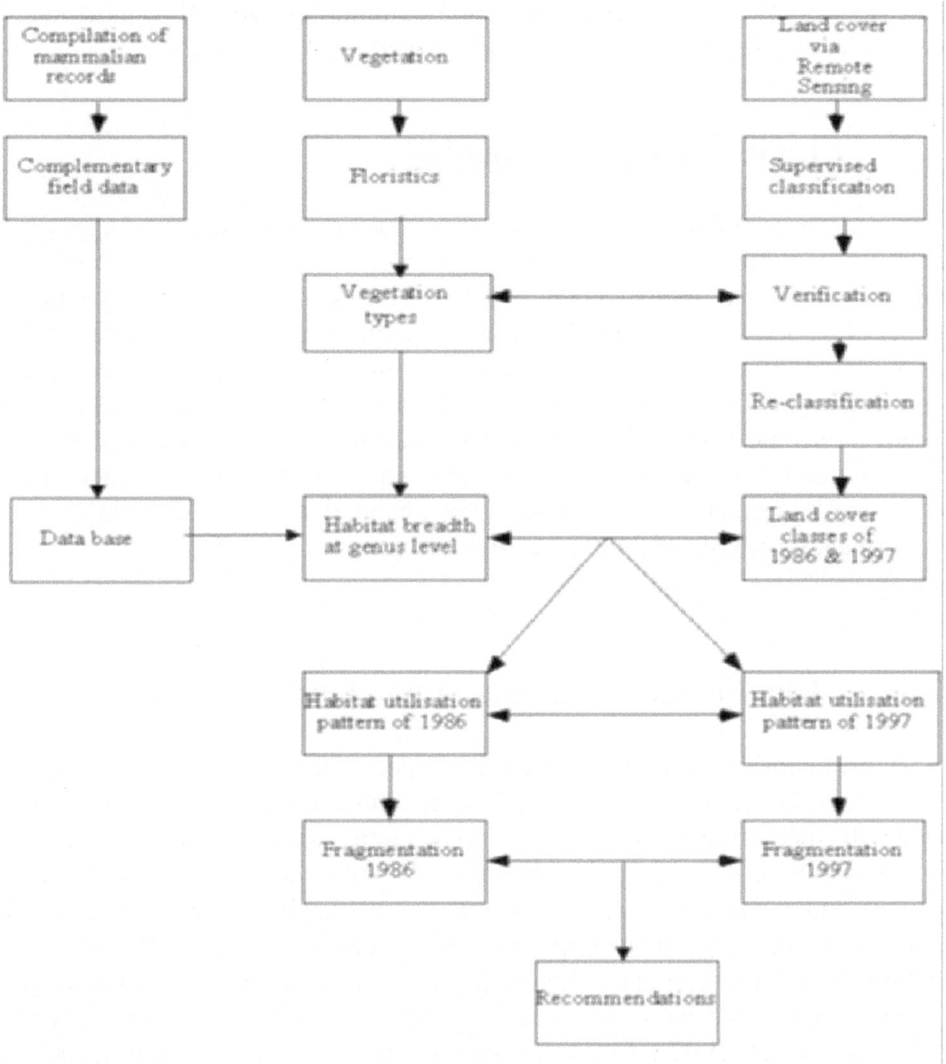

this Chapter. Three sources of information were needed to formulate sound recommendations, namely: mammalian records, vegetation data and remote sensing images.

A detailed description of all vegetation types was given elsewhere (Almeida *et al.* 1994; see also Chapter 2). The second methodology comprised the physiognomic description of all vegetation types in relation to land cover classes. Thus, floristic-physiognomic vegetation types were obtained and organized into a hierarchical scheme so that all plant communities (including the dominant plant species) and land cover types were used as a basis to create a proper habitat description *(sensu* Velázquez and Heil 1996). The resulting vegetation communities served to label land cover classes obtained from spectral classification (Figure 5.2).

2.4   Land cover pattern analysis

Land cover classes were obtained from the Landsat satellite images (50 x 50 m resolution). Two Landsat images were used for this purpose: one taken in 1986 and another taken in 1997. The identification of major land cover classes followed a standard classification procedure (Burrough 1988; Aronoff 1989). At first, the procedure included image geographical correction. Secondly, preliminary land cover classes were detected to conduct ground-verification. Over 200 sites were field cross-checked to validate land cover classes obtained from spectral classification (See also Chapter 3). Final land cover classes were obtained after re-classification for both images on basis of field data (Figure 5.2).

2.5   Linking mammal database to land cover classes

Most historical mammalian records at species level were geographically inaccurate. Because of this, records were clustered at a genera level. In turn, all species records belonging to a genus were treated as a single taxonomic group. The data was then used to build a data base of land cover classes and mammalian records. In this database, all records of mammals per land cover class were expressed as a frequency value reflecting the habitat breath of the genus (Figure 5.2). Land cover classes with a large number of records were assumed as areas of large habitat heterogeneity (Edwards *et al.* 1997). Consequently, an index of habitat heterogeneity based on mammalian record frequency values was assigned to all land cover classes.

2.6   Land use changes

Land cover classes harboring a similar number of mammalian genera were clustered as one single habitat type. This was made by re-grouping land cover classes according to substantially different threshold values of mammalian records. The threshold values to define habitat types were as follows: land cover classes including at most eight genera (as low habitat richness); land cover classes harboring from 10 to 20 genera (as medium habitat richness); and land cover classes containing more than 23 genera (as high habitat richness). This type of habitat definition was represented in maps resulting from classification using 1986 and 1997 satellite images. Once the land cover classes were converted into habitat richness classes, the statistical analysis could be carried out. To quantify landscape and habitat changes, the two maps of 1986 and 1997 were used for cross tabulation. This was processed through a Geographic Information System (IDRISI) so that number of hectare transformation per habitat richness type was obtained. Habitat fragmentation was evaluated by using a fragmentation index, i.e. by measuring the number of different classes surrounding a pixel by its neighboring pixels. The fragmentation equation is as follows:

$$F = (n - 1) / (c - 1)$$
where: n = number of different classes present
c = number of cells considered (i.e. 9).

This fragmentation index was then calculated for all pixels in both images (1986 and 1997). First the fragmentation index was applied to both images so that every pixel gets a new habitat richness value. Then the two images, reclassified according to the fragmentation index, were compared through cross tabulation. Habitat changes were calculated and interpreted as to how these affect mammalian assemblages. Finally, the results are discussed in the light of landscape changes and possibilities for conservation alternatives for the study area.

3.   RESULTS

3.1   Mammalian richness

A total of 52 mammal species within 37 genera and 16 families occur in the study area (Appendix). Approximately 90% of the terrestrial families recorded for the Valley of Mexico (quoted in Ceballos and Galindo 1984), and over 10% of all Mexican mammals are present at the Iztaccíhuatl and Popocatépetl volcanoes. Rodents were the best represented with 5 families and 21 species, followed by carnivores (4 families and 11 species), bats (2 families and 10 species) and insectivores (1 family and 4 species). This large diversity is found at a relatively small surface (ca. 0.02% of the surface of Mexico); this demonstrates the importance of this region for wildlife refuge.

3.2   Habitat breadth

A high percentage (70%) of the total amount of mammalian genera is present in most land cover classes. These classes comprise large-sized species, all carnivores and omnivores, most predators, and very mobile animals (Figure 5.3). Some exceptions are found such as *Peromyscus* (rodents) in which most species are class-specific but the genus is present in most land cover classes. On the whole, the expected results were obtained. Large mammals are found at most habitats whereas small ones tend to be more restricted than the large ones. Habitat changes, therefore, substantially affect more small-sized mammals. This can be further illustrated by the presence of at least three endemic endangered genera (*Neotomodon*, *Romerolagus* and *Cratogeomy*s). Thus, small-sized mammals and habitat-specialised mammals occur only in restricted land cover classes (e.g., *Romerolagus diazi* cf. Velázquez *et al.* 1996). The insectivore *Cryptotis* forms an exception, showing a large habitat breadth that can be explained from a historical point of view since in the past they were distributed across larger areas than the current ones (Hall and Kelson 1959).

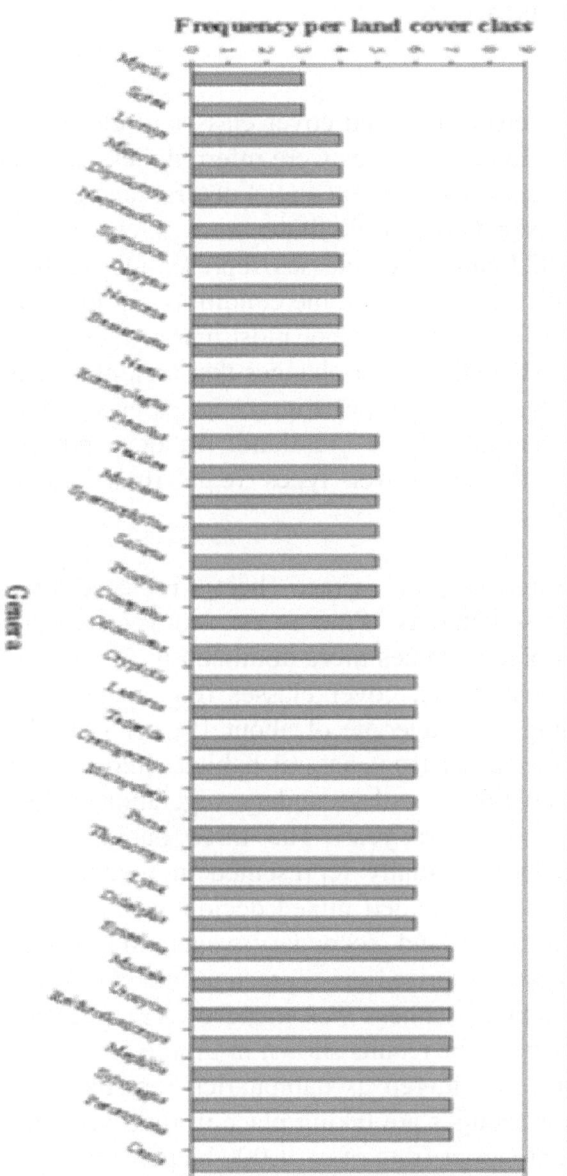

Figure 5.3. Overall, nine land cover classes were distinguished from the Landsat images of 1986 and 1997. The graph gives the land cover classes in which the genera are found. Two genera occur at three land cover classes only (Myotis and Sorex), whereas Canis is present in all of them. The rest genera occurred in four to seven land cover classes. The frequency value of the genera per land cover class was used as an index of habitat richness.

3.3   Land use changes

It was possible to detect nine land cover classes in the images of 1986 and 1997. Two major land cover types were inferred, i.e. a man-made (e.g., crops and secondary vegetation) and a (semi-) natural land cover type (e.g., fir forest, alpine grassland). Overall, these were re-classified into three habitat richness types. Land cover classes representing snow, rocks and urban settlements depict the less favorable conditions for mammals. On the contrary, most natural forest types harbor most mammalian genera. This is shown in Figure 5.4, where the major changes that took place from 1986 to 1997 are given per land cover type. Figure 5.4 shows major land use transformation from natural (alpine grassland, fir, open pine, and mixed forests) into man-made land cover types (e.g., rural settlements and secondary vegetation).

Crops have also reduced in surface, being replaced by secondary vegetation and rural settlements. Pine forest, conversely, increased its distribution range, as pine becomes more dominant in fir and mixed forest. Natural, rather than man-made, cover classes (e.g., pine, fir and mixed forest), are diminishing at an average of about 1% per year. This implies that the whole transformation from natural habitats into man-made cover types (mainly rural settlements and secondary vegetation) takes place at a rate of approximately 6% per year. The three habitat richness types previously identified were spatially represented in Figure 5.5. The class with high habitat richness, included mixed deciduous-conifer forest, open pine forest, fir with forest and secondary vegetation. The class with medium habitat richness comprised pine forest, alpine grassland and crops. The class with low habitat richness was made of rural settlements, snow and rock land cover types. The results shown in Figure 5.5 demonstrate that relatively little surface considered as habitat rich remained by 1997 (ca. 10,000 ha). The major changes are taking place in rural settlements, which are rapidly growing onto crops (rate of ca. 1,000 ha per year).

Agriculture encroachment, in turn, is depleting fir and mixed deciduous-conifer forest at a rate of ca. 500 ha/year. This explains that land cover type labeled as crop is diminishing but deforestation is still taking place, since crops are moving upwards onto forested areas (Figure 5.6).

Figure 5.4. Comparison of land use changes reflected in habitat richness for mammalian

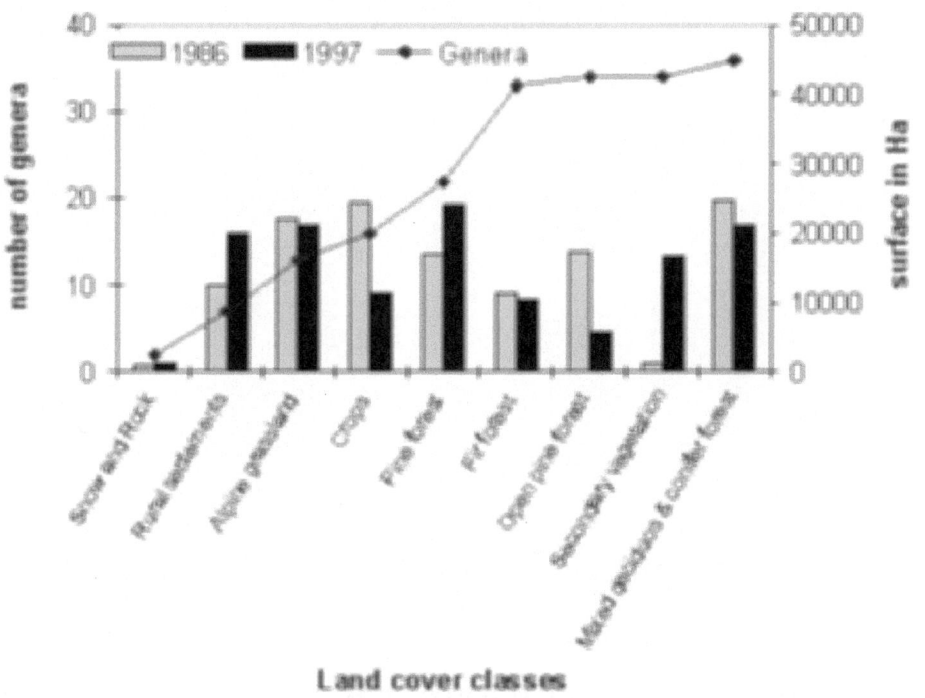

Land cover classes

genera. The line shows the accumulative mammalian genera richness across all habitat types as detected. Land cover classes including at most 10 genera were considered as low habitat richness; those including 10 to 20 genera as medium habitat richness; and those harbouring more than 20 genera as high habitat richness.

Alpine grassland is reducing its surface at a rate of about 100 ha/year mainly due to cattle grazing, fire and man-made activities, mostly promoting establishments of secondary vegetation and open pine forest (Figure 5.6). As documented by Velázquez (1993) for a nearby area, fire, livestock grazing and timbering favor the latter two land cover classes. In addition to changes in land cover classes from 1986 to 1997, large environmental degradation is taking place, mainly transforming homogeneous habitat rich areas into scattered remnants (Figure 5.7). Major changes from high to medium and medium to low habitat richness classes are happening throughout the whole area.

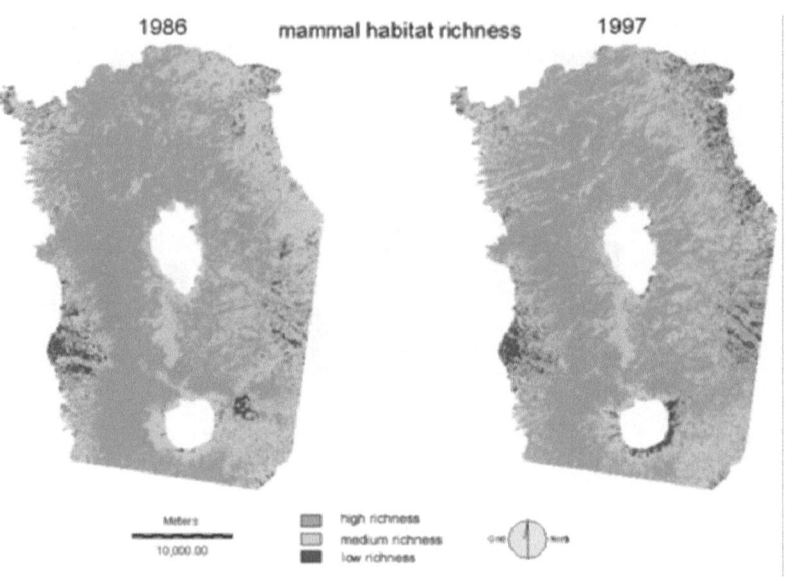

Figure 5.5. Maps showing the three habitat richness types detected in the region according to mammalian genera in 1886 and 1997. Rock above 4500 m of elevation and permanent snow land cover types were not assigned to any habitat type and are indicated by white in the images.

This reduces the resilience of the natural landscapes and increases the threats for most mammalian species since connectivity is important to fulfill daily and yearly mobility. In Figure 5.7 it can be seen that gray pixels represent those areas surrounded by pixels of the same habitat type, regardless of their richness. In other words, these areas are experiencing no fragmentation at all. Light gray pixels are areas connected by one or two pixels of the same habitat type but surrounded by pixels from other habitat types too. Most fragmentation is taking place at pixels shown in black, which are surrounded by pixels of different habitat richness. This result suggests that most of the area is experiencing large isolation of habitats. Comparing the fragmentation process between 1986 to 1997 images, it can be observed that only about 17% of the whole area is presently forming continuous habitats (gray pixels), whereas about 80% (light gray pixels) is already undergoing some form of habitat transformation.

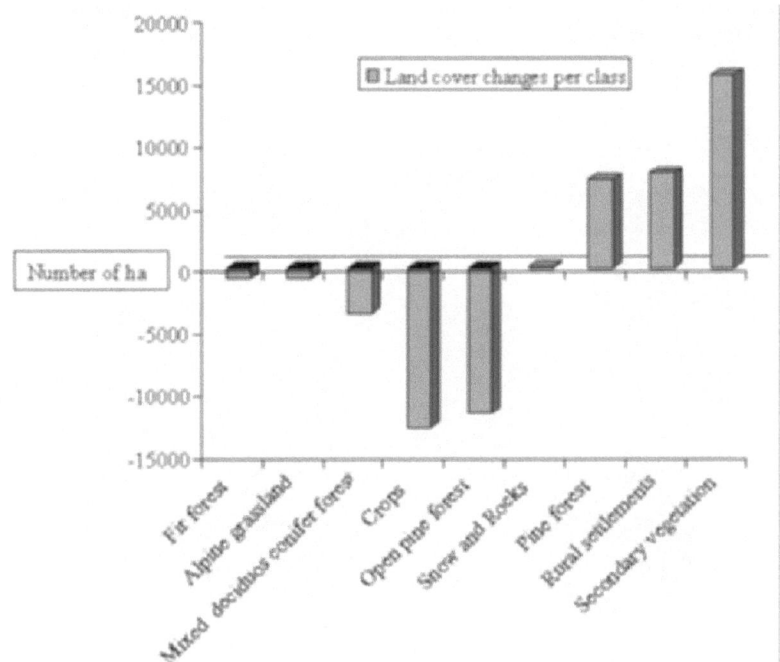

Figure 5.6. Land cover changes taking place between 1986 and 1997 among land cover classes. The classes with negative numbers are decreasing whereas the positive ones are increasing in surface.

4.   DISCUSSION AND CONCLUDING REMARKS

Although it has been proved that wildlife, especially mammals and birds, may play an important role in ecological, economic and social values, as yet little research has been done in the area to include wildlife in land use management plans (Curry-Lyndhal 1982). In our study area, most records (direct and indirect) of wild mammals are about to vanish since large land use changes are taking place as a product of drastic, non-planned, human activities (Figure 5.6). This habitat transformation may diminish the potential value of mammals comprised within the area (Wahlberg *et al.* 1996). Any land cover change impact in one way or another current mammalian assemblages. Fir forest, for example, is very important for squirrels *(Sciurus* and *Spermophylu*s), racoons *(Procyo*n), pumas *(Pum*a), and skunks *(Conepatus* and *Mephiti*s), among others.

Alpine grasslands, in turn, are a restricted habitat for the volcano mouse *(Neotomodon)*, and preferential habitat for all ground-specialized genera (e.g., *Microtus, Sorex, Mustela).*

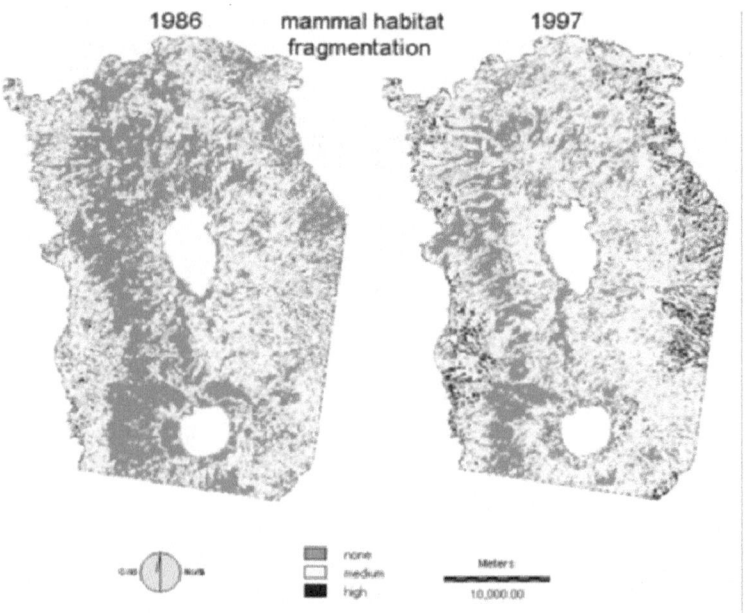

Figure 5.7. Maps depicting fragmentation of the three habitat richness types for mammals in the study area. Grey pixels represent the only areas under no threat of fragmentation.

To illustrate this further, about one fifth of the species are endemic, five are endangered and 24 subspecies are practically restricted in distribution to the area (Appendix). Clear examples of these taxa are *Romerolagus diazi* (volcano rabbit), *Cratogeomys merriami* (volcano gopher), *Neotomodon alstoni* (volcano mouse) and *Reithrodontomys chrysopsis* (volcano harvest mouse) and the subspecies *Thomomys umbrinus vulcanicus, T. u. peregrinus* (gophers), *Sorex vagrans orizabae* (shrew), *Peromyscus aztecus hylocetes* (mouse), among others (Flores and Gerez 1994). It should be mentioned that all natural land cover types provide habitat for different mammalian assemblages. For conservation purposes, however, the class medium habitat richness (especially pine forest and alpine grassland cover types with a few patches of crops) harbors five of eight endemic species represented in five genera occurring in *Romerolagus, Neotomodo*n, *Dipodomys, Liomys* and *Sigmodo*n.

The purpose of linking mammalian records to land cover classes was twofold. On the one hand, to evaluate the actual habitat trend by looking at land cover class changes and, on the other, to detect which mammalian genera distribute at similar habitat conditions. The present analysis shows which cover types may have priorities for mammal conservation. If we use mammals as umbrella taxa, holistic landscape planning efforts can be undertaken (e.g., Velázquez and Heil 1996). Abundance values, survival rates, reproduction success and other ecological attributes are the most recommended attributes for modelling wildlife conservation (Van Horne 1983; Morrison *et al.* 1992).

These sorts of data, however, may not be available in the short term whereas conservation actions are urgent. Landscape modelling for land use management, therefore, should be considered as an alternative to compromise ecologically sound landscape modelling and conservation at the Iztaccíhuatl and Popocatépetl volcanoes (e.g., Velázquez 1996; Ridgley and Heil 1998; see also Chapter 8). In this study habitat class definition was regarded as clusters of habitats that shared large affinity in two aspects. First, in their structure and composition of vegetation and second in the amount of mammalian genera present. Population and abundance data are not available for most mammalian species so that significant habitat tests cannot be conducted yet.

This task is yet to be carried out to ponder differences in habitat for mammals. Hence, the present habitat classification ought to be considered as a preliminary overview of mammal distribution. The present Chapter provides data to demonstrate the importance of the area as a unique mosaic of mammalian assemblages. The core of the data, however, came from non-updated records so that some of the species may not be present anymore. For instance, the bat *Plecotus mexicanus* and the rodents *Peromyscus maniculatus fulvus*, *P. dificilis*, *P. aztecus*, *Thomomys umbrinus vulcanicus* and *Dipodomys phillipsii* have not been recorded for the last 20 years within the study area. The lack of these records may be due to the large landscape changes described in this paper. Therefore, it is necessary to undertake a continuous monitoring of mammal species to avoid extinction of restricted and endemic taxa.

ACKNOWLEDGEMENTS

F. Cervantes-Reza allowed access to the mammal collection at the Instituto de Biología, UNAM. Funding was given from IFS (project D2540-1), UNAM-DGAPA (project ES101196) and UAM-Xochimilco.

REFERENCES

Almeida, L., Cleef, A.M., Herrera, A., Velázquez, A. and Luna, I. (1994). El zacatonal alpino del Volcán Popocatépetl, México,y su posición en las montañas tropicales de América. Phytocoenologia 22(3): 391–436.

Aronoff, S. (1989). Geographic Information Systems. A management Perspective, p 294. WDL Publications,Ottawa.

Blanco, S., Ceballos, G., Galindo, C., Maass, M., Patrón, R., Pescador, A. and Suárez, A. (1981). Ecología de la Estación Experimental Zoquiapan: descripción general, vegetación y Fauna. Cuadernos Universitarios No. 2. Universidad Autónoma de Chapingo, México, D.F.

Brown, J.H. (1995). Macroecology, p 269. Chicago University Press, Chicago.

Burrough, P.A. (1988). Principles of Geographical Information Systems and Land Resources Assessment, p 194. Clarendon Press, Oxford, UK.

Ceballos, G. and Galindo, C. (1984). Mamíferos silvestres de la Cuenca de México, p 292. Limusa, México, DF.

Chavéz, J.M. and Trigo, N. (eds.) (1996). Programa de Manejo para el Parque Nacional Iztaccíhuatl-Popocatépetl. Colección Ecología y Planeación Universidad Autonoma Metropolitana, Xochimilco.

Curry-Lindhal, K. (1982). Principios ecológicos para la conservación de los animales. In: Sioli H et al. (eds) Ecología y protección de la naturaleza; conclusiones internacionales, pp 192–242. H. Blume, Madrid, Spain.

Demant, A. (1978). Características del Eje Neovolcánico Transmexicano y sus problemas de interpretación. Revista del Instituto de Geología de la Universidad Nacional Autónoma de México 2: 172–188.

DETENAL (1983). Carta de uso del suelo y vegetación. Escala 1: 50 000.

Dirzo, R. and Gómez-Pompa, A. (1997). Reservas de la Biosfera y otras áreas naturales ptotegidas de México, p 159. CONABIO, Mexico.

Edwards, P.J., May, R.M. and Webb, N.R. (1997). Large-Scale Ecology and Conservation Biology, p 375. Blackwell Science, Oxford, UK.

Ehrlich, P.R. and Wilson, E.O. (1991). Biodiversity studies: science and policy. Science 253: 758–762.

Ferrari-Pérez, F. (1886). Catologue of animals collected by the Geographical and Exploring Commission of the Republic of Mexico. Proceedings of the US Natural Museum 9: 125–199.

Flores, V.O. and Gerez, P. (1994). Biodiversidad y conservación en México: vertebrados, vegetación y uso del suelo, p 439. UNAM-CONABIO, México, DF.

García, E. (1981). Modificaciones al sistema de clasificación de Koeppen, p 246, 2nda edición. Instituto de Geografía, Universidad Nacional Autónoma de México. México, DF.

Hall, E.R. and Kelson,, K.R. (1959). The Mammals of North America. The Ronald Press Company, New York.

Hamilton-Smith, C. (1839). The Naturalist's Library, p 223. Jardine.

Hoth, J., Velázquez, A., Romero, F.J., León, L., Aranda, M. and Bell, D.J. (1987). The volcano rabbit – a shrinking distribution and a threatened habitat. Oryx 21: 85–91.

Lugo, H.J. (1984). Geomorfología del Sur de la Cuenca de México. p 95. Instituto de Geografía, Universidad Nacional Autónoma de México. Serie Varia T. 1. Número. 8. México, DF.

Maza, R. and Soberón, J. (1998). Morphological grouping of Mexican butterflies in relation to habitat association. Biodiversity and conservation 7(1): 927–944.

Merriam, C.H. (1895). Synopsis of the American shrews of the genus *Sorex*. North American Fauna 10: 57–124.

Merriam, C.H. (1898). A new genus *(Neotomodo*n) and three new species of murine rodents from The moutains of southern Mexico. Proceedings of the Biological society of Washington 12: 127–129.

Mittermeier, R.A. and Mittermeier, G.C. (1998). Megadiversidad, p 498. CEMEX, Mexico City, Mexico.

Morrison, M.L., Marcot, B.G. and Mannan, W.R. (1992). Wildlife–Habitat Relationships, p 343. The University of Wisconsin Press, Madison, Wisconsin.

Mueller-Dombois, D. and Ellenberg, H. (1974). Aims and Methods of Vegetation Ecology. John Wiley and Sons, New York.

Myers, N. (1993). Questions of mass extinction. Biodiversity and Conservation 2(1): 2–17.

Nelson, E.W. and Goldman, E.A. (1934). Pocket gophers of the genus *Thomomys* of Mexican mainland and bordering territory. Journal of Mammalogy 15: 105–124.

Ridgley, M. and Heil, G.W. (1998). Multicriterion planning of protected-area buffer zones: an application to Mexico's Izta-Popo National Park. In: Beinat E and Nijkamp P (eds) Multicriteria Analysis for Land-Use Management, pp 293–311. Kluwer Academic Publishers, Dordrecht, The Netherlands.

Rojas, M.P. (1951). Estudio biológico del conejo de los volcanes (género *Romerolagus*) (Mammalia: Lago-morpha). Tesis profesional. Fac. Ciencias, UNAM, México, DF.

Santillán, A.S. (1979). Distribucion altitudinal de roedores en el Campo Experimental 'San Juan Tetla', Estado de Puebla, México. Tesis profesional. UAEM, México, DF.

Saussure, M.H. (1860). Note sur quelques mammifères du Mexique. Rev. Mag. Zool. (2) 12: 3–11; Feb. 53–57; Mar. 97–110; Jun. 241–254; Jul. 281–293; Sep. 377–383; Oct. 425–431 and Nov. 479–494.

Schwartz, M.W. (1993). Modelling effects of habitat fragmentation on the ability of trees to respond to climatic warming. Biodiversity and Conservation 2(1): 51–61.

Shilling, F. (1997). Do habitat conservation plans protect endangered species? Science 276: 1662–1663.

Sutherland, J.W. (1996). Ecological Census Techniques, p 336. Cambridge University Press, Cambridge.

Van Horne, B. (1983). Density as misleading indicator of habitat richness. Journal of Wildlife Management 47(4): 893–901.

Velázquez, A. (1993). Landscape Ecology of Tláloc and Pelado Volcanoes, p 151. ITC Publication No. 16, México.

Velázquez, A. (1996). Biodiversidad a la carta. ICYT, Información Científica y Tecnológica. CONACYT 3 98–101.

Velázquez, A. and Heil, G.W. (1996). Habitat analysis of the volcano rabbit *(Romerolagus diazi)* by different statistical methods. Journal of Applied Ecology 2: 193–102.

Velázquez, A and Romero, F.J. (1999). Biodiversidad de la región de montaña del sur de la Cuenca de México, p 358. UAM-CORENA, México.

Velázquez, A., Romero, F.J. and López-Paniagua, J. (1996) Ecología y conservación del zacatuche. Publicaciones científicas, Fondo de Cultura Económica, México, D.F.

Velázquez, A., Bocco, G. and Torres, A. (2001). Turning scientific approaches into practical conservation actions: the case of Comunidad Indígena de Nuevo San Juan Parangaricutiro, México. Environmental Management 3: 21–32.

Villa-Ramírez, B. (1953). Mamíferos silvestres del Valle de México. Anales Inst. Biol. Univ. Nal. Autón. Mex. 23: 269–242.

Wahlberg, N., Moilanen, A. and Hanski, I. (1996). Predicting the occurrence of endangered species in fragmented landscapes. Science 273: 1536–1538.

APPENDIX 1

Taxonomic list of all mammals reported in the study area from 1839 up to 1997. A total of 52 mammal species within 37 genera and 16 families were included in the analysis. A column indicating the year of the first record per taxa is given. Endemicity is indicated in the last column.

TAXA	First record	Endemic
DIDELPHOIDIA		
Didelphidae		
Didelphinae		
Didelphis virginiana californica	1973	-
INSECTIVORA		
Soricidae		
Sorex vagrans orizabae	1944	Subspecies
Sorex saussurei saussurei	1959	Subspecies
Sorex oreopolus ventralis	1972	Subspecies
Cryptotis goldmani alticola	1895	Subspecies
CHIROPTERA		
Vespertilionidae		
Vespertilioninae		

Myotis californicus mexicanus	1972	-
Myotis velifer velifer	1944	-
Myotis volans amotus	1980	Subspecies
Myotis thysanodes aztecus	1972	-
Eptesicus fuscus miradonensis	1944	-
Lasiurus cinereus cinereus	1981	-
Idionycteris phyllotis	1984	-
Plecotus mexicanus	1944	Species
Molossidae		
Tadarida brasiliensis mexicana	1984	-
Molossus molossus aztecus	1860	-
XENARTHRA Dasypodidae Dasypodinae		
Dasypus novemcinctus mexicana	1981	Subspecies
LAGOMORPHA		
Leporidae		
Romerolagus diazi	1893	Genus
Sylvilagus floridanus orizabae	1909	Subspecies
Sylvilagus cunucularius cunicularius	1909	Species
RODENTIA		
Sciuridae		
Sciurinae		
Spermophylus mexicanus mexicanus	1944	Subspecies
Spermophylus variegatus variegatus	1938	Subspecies
Sciurus aureogaster nigrescens	1923	-
Geomyidae		
Geomyinae		
Thomomys umbrinus aff. peregrinus	1954	Subspecies
Thomomys umbrinus vulcanius	1934	Subspecies
Cratogeomys merriami merriami	1968	Genus
Heteromyidae		
Dipodomyinae		
Dipodomys phillipsii	1932	Species

Heteromyinae		
Liomys irroratus alleni	1973	Species
Muridae		
Sigmodontinae		
Neotoma mexicana torquata	1910	Subspecies
Peromyscus difficilis felipensis	1909	Subspecies
Peromyscus truei gratus	1974	-
Peromyscus aztecus hylocetes	1909	Subspecies
Peromyscus maniculatus fulvus	1909	Subspecies
Peromyscus maniculatus labecula	1944	Subspecies
Peromyscus melanotis	1909	Species
Neotomodon alstoni alstoni	1898	Species
Reithrodontomys megalotis saturatus	1943	Subspecies
Reithrodontomys chrysopsis chrysopsis	1900	Species
Reithrodontomys sumuchrasti sumichrasti	1984	Subspecies
Sigmodon leucotis leucotis	1944	Species
Arvicolidae		
Microtus mexicanus mexicanus	1944	Subspecies
CARNIVORA		
Canidae		
Canis latrans cagotis	1839	Subspecies
Urocyon cinereoargenteus nigrirostris	1979	Subspecies
Procyonidae		
Bassariscinae		
Bassariscus astutus astutus	1984	-
Procyoninae		
Procyon lotor hernandezii	1984	-
Nasua narica molaris	1987	-
Mustelidae		
Mustelinae		
Mustela frenata perotae	1944	Subspecies
Melinae		
Taxidea taxus berlandieri	1979	-
Mephitinae		

Mephitis macroura macroura	1901	-
Conepatus mesoleucus mesoleucus	1981	-
Felidae		
Felinae		
Puma concolor azteca	1984	-
Lynx rufus escuinape	1979	Subspecies
ARTIODACTYLA		
Cervidae		
Odocoileus virginianus	1944	Subspecies

Chapter 6

Dynamics of plant communities in the Iztaccíhuatl - Popocatépetl National Park

Gerrit W. Heil
Department of Plant Ecology, Faculty of Biology, Utrecht University

Key words: spatially explicit modelling, community assembly, secondary vegetation succession, land use, land cover

Abstract: Dynamics of plant communities occur over a wide range of spatial and temporal scales. The result is that plant communities are patchy on virtually every level of space and time. Chaotic fluctuations, such as disturbances, interrupt succession processes in plant communities, which makes it often difficult to understand how they function. Moving from population dynamics to the level of the community involves a risk of generating systems too complex, e.g. by using mechanistic individual-based models, to provide insight into community assembly processes. In this chapter the combination of a deterministic and a stochastic Markov model on community assembly is discussed. The aim is to discuss the potential of such a modelling approach, and to apply it to real-life plant community dynamics of the Iztac-Popo National Park in Mexico. For this, remotely sensed data in combination with field data are used to develop a community assembly model in order to answer the question of how spatially localized disturbances may have consequences for plant community assembly. The results show that interconnections and feedbacks between two main processes, i.e. on one hand community assembly and on the other hand disturbances, can become visible. Additionally, as demonstrated by the results, this model offers the potential to analyze the impact of future ecological restoration management in the study area.

G.W. Heil et al. (eds.), Ecology and Man in Mexico's Central Volcanoes Area, 125–145.
© *2003 Springer Science+Business Media Dordrecht*

1.   INTRODUCTION

Community ecology is concerned with the properties of sets of species at given locations in time and space. There are three major theories of community assembly, i.e. the deterministic, the stochastic and alternative stable states (ASS). In the deterministic model, a community is seen as the inevitable consequence of physical and biotic factors (Clements 1916). In the stochastic model the community composition and structure is essentially a random process (Van der Maarel and Sykes 1993). The ASS theory is intermediate between the first two. Over the years, there has been an increasing focus on the use of individual based models in ecology, as documented by various authors (e.g. Huston *et al.* 1988; Hogeweg & Hesper 1990; DeAngelis and Gross 1992; Bart 1995). Huisman and Weissing (1999) discussed theories developed on competition within communities and on the (limited) predictive capabilities of such mechanistic competition models. Huisman and Weissing (1999) showed the importance of chaotic fluctuations in species abundance's. Their model was able to reproduce the overall behaviour of the observed process quite well, suggesting that a stochastic approach can help to gain insight into the functioning of community dynamics.

The importance of disturbance in community dynamics has only relatively recently been recognised (White and Pickett 1985, Chesson and Huntly 1997, White and Jentsch 2001). Chaotic disturbances interrupt succession processes in plant communities, which makes it often difficult to understand how they function (Hobbs 1998). Consequently, the end-point of a successional process is not a predictable uniform outcome; on the contrary, several community states are possible depending on (a)biotic conditions (Noble and Slayter 1980; Hobbs 1994). These multiple community states may be stable for long periods of time and distinct thresholds may exist which limit the transition from one state to another. Anand and Orlóci (1997) pointed out that community dynamics in the early state are linear, but then break down into a 'noisy' state. They showed that adding small amounts of quasi-random perturbation to a simple linear model could turn the normally well-behaved dynamics into a deterministically chaotic one. A key to improve the predictive capabilities of such stochastic models might be to switch from dynamics describing the flux of individuals to the process of community assembly involving a series of filters, which sieve species out of the regional pool (Diaz *et al.* 1998;

Diaz *et al.* 1999). There is a strong move away from working at the species level towards using functional groups of species or guilds (Grime 1979; Hobbs 1997; Wilson 1999). However, moving from population dynamics up to the level of the community, including all processes that comprise the different scales, there is a risk of generating dynamical systems too complex to provide insight into important community processes at a coarse scale of a landscape. Consequently, scale is of fundamental importance for testing such an approach, i.e. the explanation of any system depends on the spatial and temporal scale chosen. Each level of scale has its own focal point that happens with the processes at that particular level (Fig 6.1).

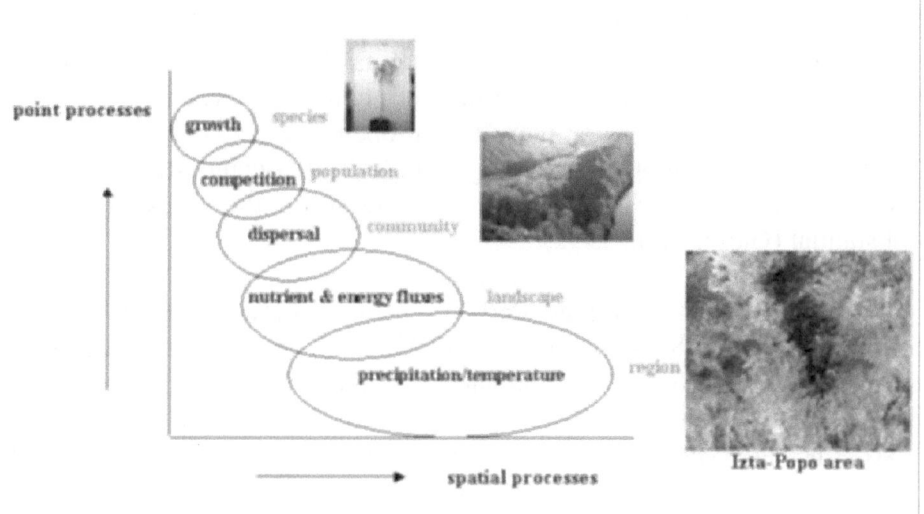

Figure 6.1. Scaling of processes

Law (1999) suggested that it could be appropriate to think eventually in terms of a filtering process with the state variable $P(C_i,t)$ being the probability that the community is in state C_i at time t. The probability per time unit of moving from state C_i to C_j, $w(C_j|C_i)$ can be seen as a stochastic Markov process depending on the probability per time unit that each species will arrive at the site, and the set of resident species which results from each arrival. In this Chapter the combination of a deterministic and a stochastic model on community assembly will be discussed. In the model, topography (slope and elevation) and distance to roads are used as stochastic filters. As an example of how deterministic and stochastic processes might interact in community assembly, the model has been applied on the plant community dynamics of the "Iztac-Popo" National Park.

2.   METHODS

2.1   Research area

Close to Mexico City, the snow-capped volcanoes of Iztaccíhuatl and Popocatépetl, part of the southernmost stretch of the Sierra Nevada, rise to elevations of well over 5,400 m and form the backbone of what is commonly referred to as Iztac-Popo National Park. Extensive forestry in the region mainly concerns cutting and collecting wood, cutting tree branches for torches or for utensils for open-fire cooking, collection of mushrooms, and hunting. Although these (often-clandestine) activities seem to be small-scale, their adverse effects on the forest have been substantial (García *et al.* 1992).

The Iztac-Popo National Park is located between 18°59'00" and 19°16'25" N and 98°34'54" and 98° 42'08" W. The position of the area is shown in Figure 6.2. The boundaries of the Iztac-Popo National Park follow the 3,600 m altitude topographic line. The surface of the park is 25,679 hectares. From the constitution of the park in 1935 until 1948, the 3,000 m topographic line, the area occupying 59,913 ha, delineated the park boundaries. Proposals have been made to lower the borders of the park to 2,500 masl; which would imply an area of 118,792 ha (Chávez and Trigo 1996). This proposal includes a management plan that consists of conservation and management, public use and recreation and administration. As part of this plan, an analysis has been carried out for designing buffer zones around protected areas in Mexico's Iztac-Popo National Park (Ridgley and Heil, 1998).

2.2   Vegetation

The . vegetation in the Iztac-Popo Park varies greatly along with altitudinal and microclimatic changes across the slopes. The tree line at this latitude is around 4,000 m (Velázquez *et al.* 2001). The vegetation that can be found in the area can be characterized roughly in the following types:

- *Abies* forest: high evergreen forest with *Abies religiosa* as dominant tree species, although other conifer tree species can be present.
- *Pinus* forest: more open than *Abies*, evergreen forest mostly with *Pinus hartwegii* as the dominant species.
- Mixed coniferous forest with *Abies* and *Pinus spp*.
- Open coniferous forest and secondary communities that develop where original forest vegetation is destroyed by cutting, grazing or fire.
- Semi-natural grasslands, homogeneous communities of bushy graminae of 60-120 cm height at 2,500-4,300 m, and natural alpine grasslands, paramos or alpine zacatonals: tropical bunch tussock grasslands from 4,300-4,500 m.
- Bare soil or rock: ground covered with hardly any vegetation

2.3   Classification

Satellite images were used for creating land cover maps through a supervised classification (see also Chapter 2). For this, Landsat-Thematic Mapper (TM) images from January 1986 and February 1997 of the Iztac-Popo area were processed according to standard procedures (Jensen, 1986; Belward and Valenzuela, 1991; Matson and Ustin, 1991; Lilesand and Kiefer, 1994). The ground resolution of the Landsat TM images is 30 x 30m. For practical reasons of data analysis and processing, but also to facilitate overlaying with existing maps such as digital road maps and Digital Elevation Map (DEM), the TM images have been resampled to a 50 x 50 m resolution. Principally, this will hardly affect the classification of the TM-images, because for our purpose the classification should lead to a classification of the dominant plant communities. For the supervised classification of the TM images ground control points were gathered at 223 different locations in the whole area of the Iztac-Popo National Park and buffer zones. For all the ground control points the georeferenced position was obtained with a Global Positioning System (GPS) and data annotated about the dominant vegetation including diameter at breast height (DBH) of the trees and the percentage of canopy cover of all plant species.

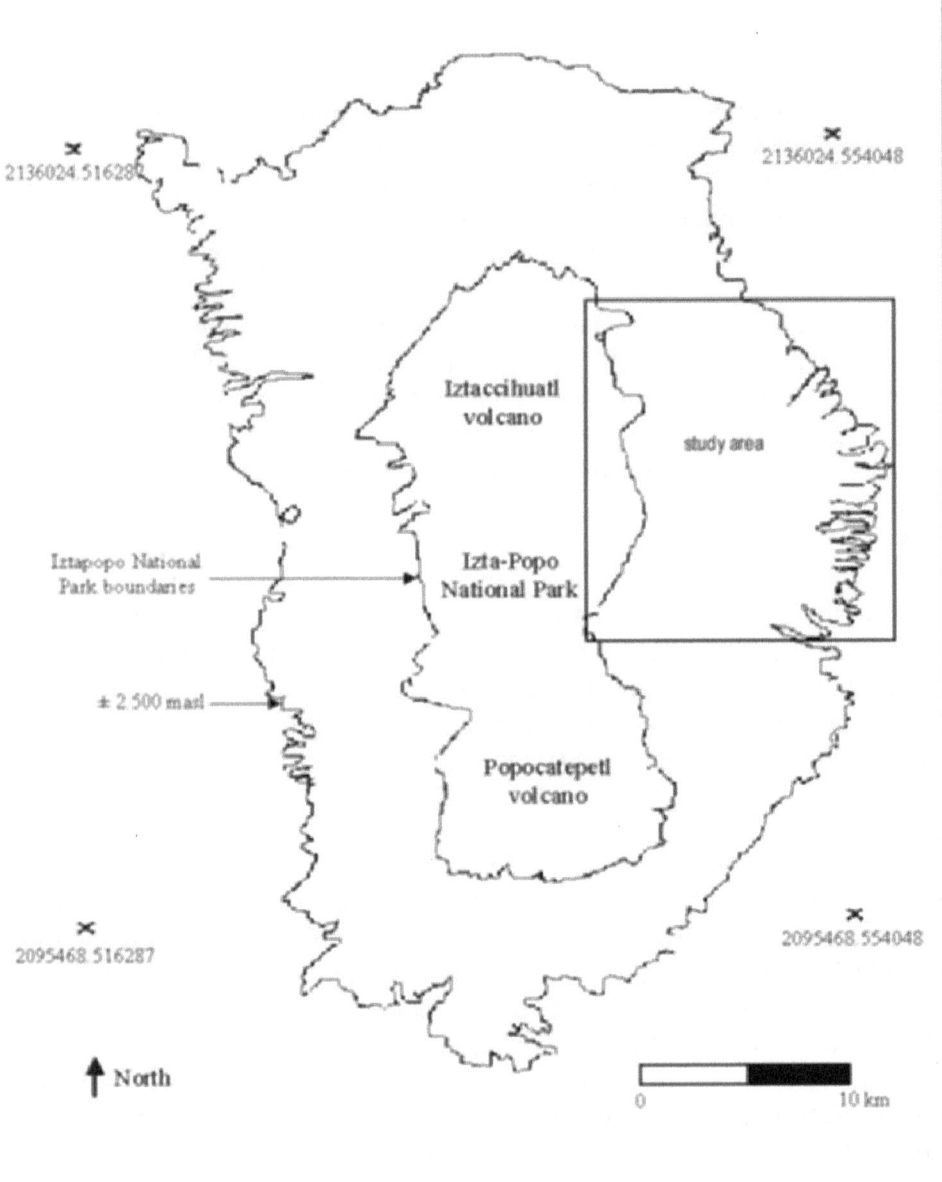

Figure 6.2. Location of the Iztac-Popo National Park, its environment and the modelling site in the eastern slope of the Iztaccíhuatl volcano.

2.4   Modelling

A matrix was used to summarize community transitions. Matrix modelling is a widely used method in which the transitions among communities are used to build a simplified description of the relationships (Van Hulst 1979; Caswell 1989; Orlóci *et al.* 1993). A model must be constructed before it can be analysed. Model construction requires not only knowledge of the analytical tools that will eventually be applied to the model, but also concern for the necessary data and for the manipulations needed to transform the data to the required form. If we consider the dynamics of the plant communities as a cycle of specific phases, then the different plant communities are part of a permanent cycle of a species pool comprising the different community types (Rosenzweig 1995). Each community can be invaded by species from one or more of the other communities.

Consequently, the union of these communities could constitute a permanent cycle of this species pool. The land cover of part of the maps of 1986 and 1997 were used for cross tabulation to quantify land cover changes. The transitions observed during 1986–1997 were not unusual because of the climate variation, and additionally in that period important events were left out. Another part of the map, i.e. the the eastern slope of the Iztaccíhuatl, was withheld for validating the model. On basis of cross tabulation, a matrix has been created which has been used as basis for the computation of the community dynamics (Figure 6.3).

This matrix was constructed as follows.
1. On basis of the cross tabulation the projection interval has been determined to be from 1986 to 1997.
2. Depending on elevation, each community can shift with a certain probability into one of the other communities.
3. Then each probability is labelled by a coefficient; the coefficient a_{ij} on the transition from n_i to n_j gives the total cover of community *i* at time $t+1$ per community in stage *j* at time *t*. So that

$$n_i(t+1) = \sum_{j=1}^{s} a_{ij} n_j(t) \qquad \text{(equation 1)}$$

These coefficients are the transition probabilities taken from the cross tabulation of the maps of Iztac-Popo National Park and buffer zones (Table 6.1).

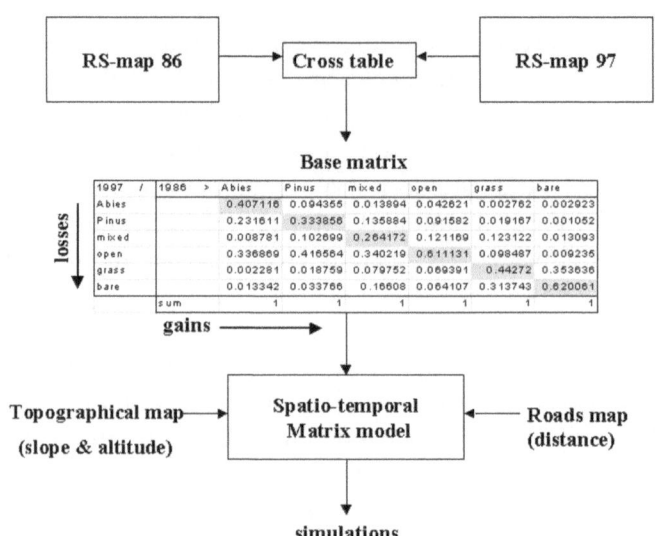

Figure 6.3. Diagram of the modelling procedure. The cross table of the two maps is used as input for the base matrix, which is used in the Markov model. The topographic map and roads map are used as distance operators for a disturbance filter

Equation 1 can be re-written as the matrix projection equation:

$$n_i\,(t + 1) = An_i(t) \qquad \text{(equation 2)}$$

Where: n_i = cover of plant community i and A = the projection matrix of plant community i

Disturbance does not take place homogeneously over the area, but is merely a filtering process depending on the distance from disturbances.

Table 6.1. Cross table of the two maps 1986 and 1997, which is used as model input. Losses show the ratio of how much of a certain community is lost towards its own or other communities, and gains show how much is gained from other communities (see also text).

Losses gains	Abies	Pinus	mixed	open	grass	bare
Abies	0.40711598	0.094355	0.013894	0.042621	0.002762	0.002923
Pinus	0.23161136	0.333856	0.135884	0.091582	0.019167	0.001052
Mixed	0.00878093	0.102699	0.264172	0.121169	0.123122	0.013093
Open	0.33686851	0.416564	0.340219	0.611131	0.098487	0.009235
Grass	0.00228076	0.018759	0.079752	0.069391	0.44272	0.353636
Bare	0.01334246	0.033766	0.16608	0.064107	0.313743	0.620061
sum	1	1	1	1	1	1

Since spatially localized disturbances of the plant communities by human activities were considered as the main disturbances, distance from roads has been used as proximate for the distance of the impact from disturbances (See also Chapter 3). It was assumed that people are confined within a certain distance that they are willing to travel for obtaining wood. In addition, it was assumed that also slopes would hinder people to cross the landscape.

Thus, both distance and slope determine the intensity of disturbance, i.e. the further away from a road and the steeper a slope the fewer the disturbances. The location of the two criteria was obtained from a digitised road map and a digital elevation model, respectively (Figure 6.4). Although different scenarios can be applied, it is assumed that the behaviour of people does not vary with resource scarcity, resource monetary value or human population size. The maps of the two criteria were used in a spatial modelling environment to calculate the probability for changes. The transitions from the matrix are converted in such way that calculations could be executed in time steps of one year.

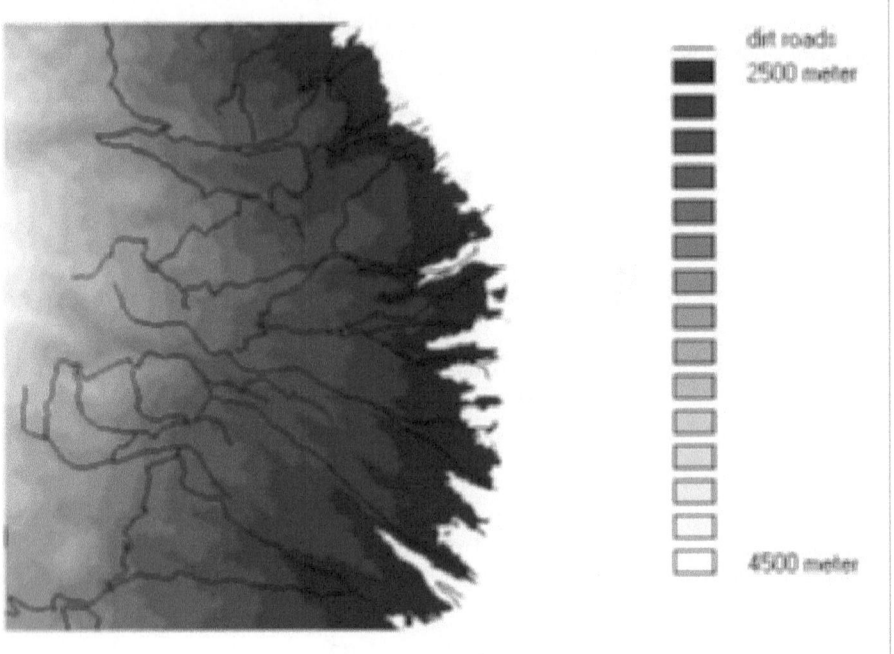

Figure 6.4. Digital elevation model (DEM) based on the topographical map and map of the roads in the eastern slope of the Iztaccíhuatl.

These transitions have been implemented in the PCRaster software (Van Deursen, 1995). The architecture of the system permits the integration of environmental modelling functions, such as transitions, with classical Geographical Information System (GIS) functions (e.g. Van Deursen and Heil 1993; Heil and Van Deursen 1996). For the numerical implementation of the simulation model it was determined that the calculated map with the actual community was classified according the dominance of the community composition, viz. an iterative procedure determined which plant community was the dominant. For initialisation of the model part of the land cover map of 1986 has been used, i.e. the eastern slope of the Iztaccíhuatl (Figure 6.2). Next the model has been used for different simulations in which the impact from the distance to roads and the angle of the slope varied. For testing the predictive capabilities of the model the simulation results have been compared against part of the land cover map of 1997 that comprises the same area, i.e. to test whether the model could produce similar patterns as found in the land cover map of 1997. The comparison has been executed through a geostatistical test according to the Kappa Index of Agreement (KIA) (Burrough 1986; Eastman 1993).

3.   RESULTS

Figure 6.5 shows the land cover maps of 1986 and 1997 of the validation site in the eastern slope of the Iztaccíhuatl. The result of the cross tabulation between the two maps of the area, excluding the validation site, is shown in the matrix of Table 6.1. In this matrix, losses show the ratio of how much of a certain community is lost towards its own or other communities and gains show the ratio of how much is gained from other communities. The numbers in the vertical and horizontal are similar and reflect the different communities; e.g. *Abies* loose 0.231611 towards Pinus and gains 0.094355 from Pinus. The matrix in Table 6.1 shows that relatively large amounts of the *Abies* forest, of the *Pinus* forest as well as of the mixed forest is transformed into open forest, viz. approximately 33%, 42% and 34%, respectively. Open forest mainly remains open forest, while grassland is to a large amount converted into bare soil, and bare soil remains bare soil for 62% and reverts for 35% to grassland. From these results, it is obvious that as the area under pressure continues to increase every year, the resulting opening of large forested areas, soil modification, and ensuing erosion makes it almost impossible for forest recovery.

The model has been applied to the land cover map of 1986 to simulate assembly dynamics of the plant communities towards the year 1997. As explained in the methods, different simulation runs have been executed which differed in the impact from maximum distance and maximum slope angle as constraint for disturbance. However, the underlying topographic map stayed the same during all runs. After each run the resulting map has been tested for comparison with the map of 1997. By doing so, a kind of optimisation could be performed from which conclusions can be drawn on how far and under which maximum angle of a slope people are willing to go to get their resources from the forest. Considering the fundamental assumptions about the transitions and the constraints of the distance and the slope, the best simulation results is shown in Figure 6.5.

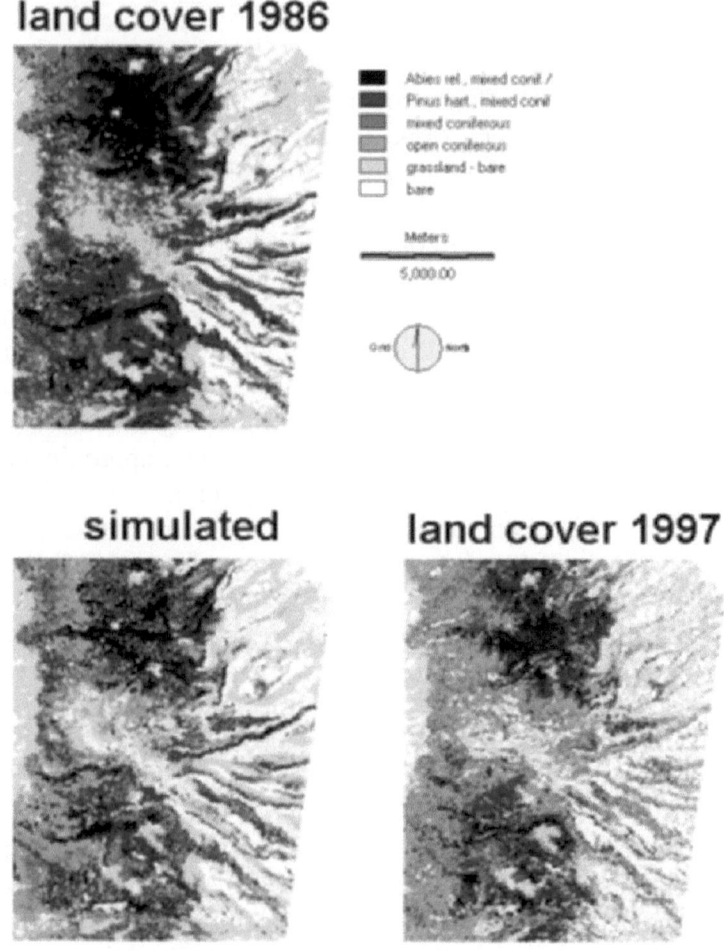

Figure 6.5. Maps of the land cover in 1986 and 1997, and resulting map of the simulation. The different grey tones represent different land cover types dominated either by *Abies religiosa*, *Pinus hartwegii*, mixed coniferous forest of *Abies* and *Pinus*, open secondary forest, grassland or bare soil.

The result of the geo-statistical tests produces a correlation coefficient that ranges from 0.0 indicating no correlation to 1.0 indicating perfect correlation. The Kappa Index of Agreement (KIA) between the simulation result and the land cover map of 1997 is 0.45, which is significant (P < 0.05) but rather low. However, the overall changes of the plant communities in these simulations into the other plant communities are in agreement and in the same direction as the actual changes found in the land cover maps from 1986 to 1997. In addition, when the *Abies*, *Pinus* and mixed forest classes are combined into one class, the simulated output match the land cover map of 1997 fairly well (KIA = 0.71). Similarly, with the land cover map of 1997, the simulated map still has a rather high correlation coefficient with the land cover map of 1986 (KIA ≈ 0.6). The time series of the simulation show that the changes are towards a dominance of open forest and grassland (Figure 6.6).

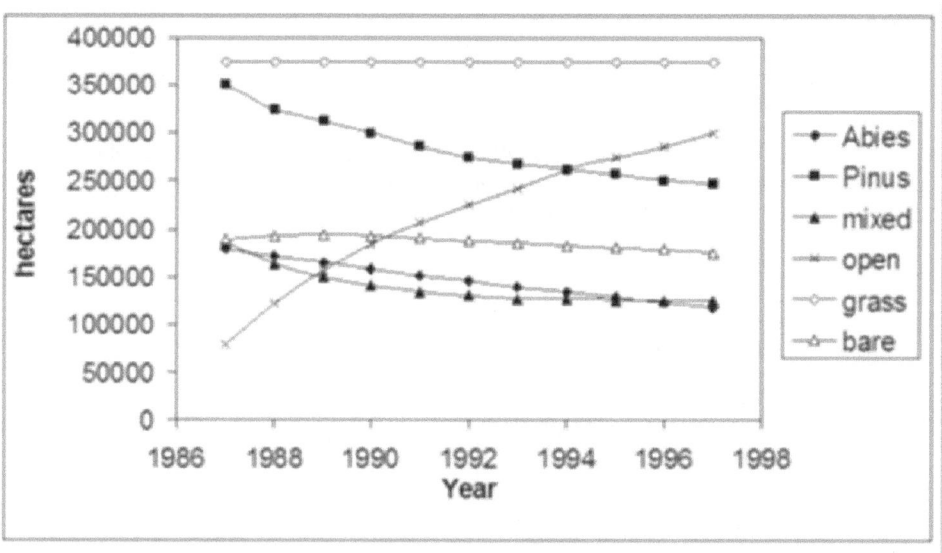

Figure 6.6. Time series of the simulations of the model including a spatially explicit disturbance filter as caused by human impact depending on distance from roads and slope angle.

When only the basic transitions in the simulation model (i.e. mean field simulation) are applied, i.e. without the simulated effects in relation to distance to roads and slope angle, the similarity with the land cover map of 1997 is not significant anymore (KIA < 0.25), and the pattern of the vegetation changes towards uniform end states, strongly dominated by continuous patches open forest, bare soil and grassland (Figure 6.7).

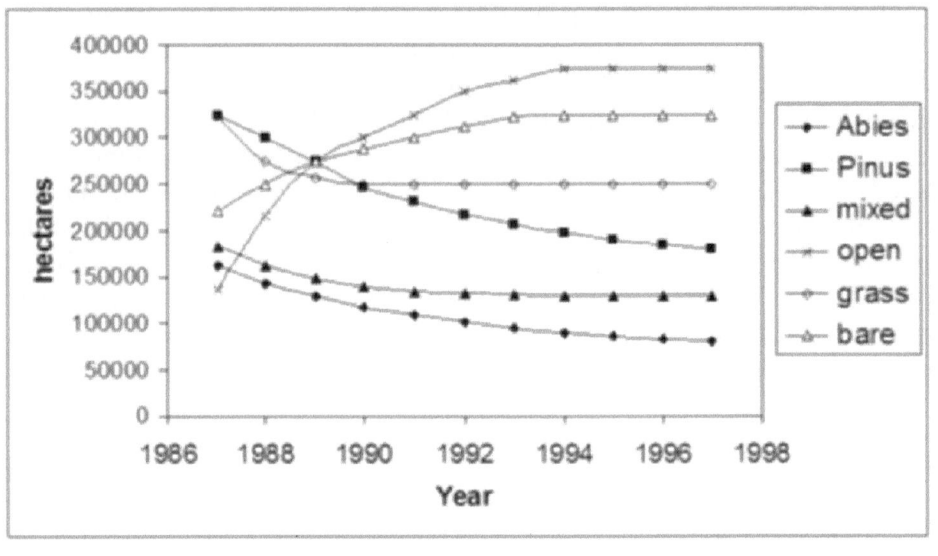

Figure 6.7. Time series of the simulations of the model without spatially explicit disturbances (mean field simulation) by human impact, i.e. human impact is evenly distributed over the modelling area

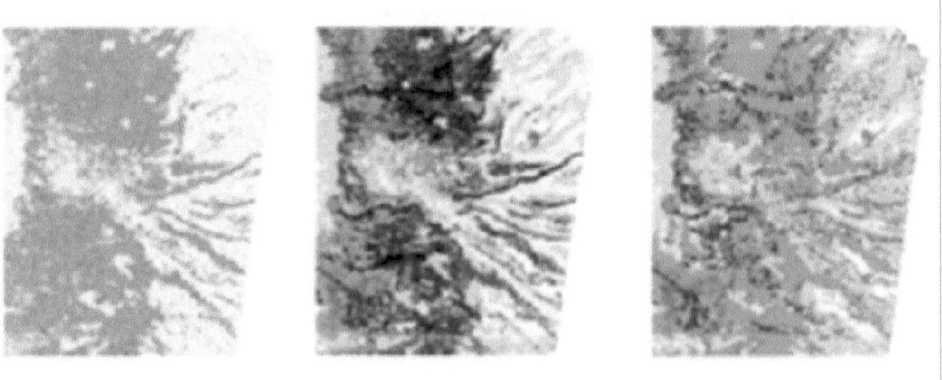

Figure 6.8. Results of basic matrix simulation after 11 years (left), and spatially explicit modelling after 11 years (middle), and 50 years (right), respectively

Even when the results of the long term simulation (50-year), including the simulated effects related to roads and slope angle, are compared with the land cover map of 1997, this "long term" map has a spatial pattern that still is strongly related to the pattern of the land cover map of 1997 (KIA = 0.45) (Figure 6.8).

4. DISCUSSION AND CONCLUSIONS

The basic difference between the modelling approach used in this study, being a non-mechanistic empirical model based on chance processes, and the mechanistic non-stochastic models on which ecological theory is mostly based, is not to uncover underlying mechanisms of community assembly but to analyze the adverse effects of relatively small-scale filter processes which may seem to be not substantial on the community assembly process. Hence, this study is much about what to leave out for modelling at the plant community level. Clearly, scale effects our perception of community pattern, process and mechanism (Anand 1994).

From the work of Law (1999), the concept has been used to switch from dynamics describing the flux of individuals to dynamics describing the community assembly involving a series of filters. Taking the latter into account it would be opportune to think in terms of a combination of deterministic and stochastic processes, in which the assembly process can be seen as a Markov process. Considering the dynamics of the plant communities as a cycle of specific phases, the different plant communities are part of a permanent cycle of a species pool comprising the different community types. This union of communities was used for the construction of a spatio-temporal simulation model. As a result, principal features of interacting (spatial) processes of community assembly and land use could be investigated. The results of a field study from the Iztac-Popo National Park and buffer areas were used to calibrate the model and to compare the results of the simulations with actual data (See also Chapters 2 and 3).

The results of the simulation model agree fairly well with the land cover map of 1997. However, using only the mean field transitions of land cover change from 1986 to 1997 for the simulation did agree to a much lesser extent, and assembled the land cover pattern uniformly towards homogeneous end states. Including distance from roads and slope angles as proximate for the intensity of disturbances by human impact did significantly improve the results. From these simulations it could be concluded that there is a maximum distance of 950 meters, and a maximum of 35 degrees of the slopes that people are willing to transverse the area for collecting wood resources. This way of modelling suggests that the behaviour of people is constant, and does not change under different conditions. Of course this is not true, because presumably people have particular preferences with respect to resources, and there is little reason to believe that people have no access to the rest of the area. However, for data availability reasons only, the criteria for accessing are based on a digital road and slope map.

Although the simulation model could not be validated firmly (*cf.* Jörgensen 1988), the model has shown to be rather persuasive and is transparent in its simplicity. It is obvious that in a model like this one, complicated population dynamic processes are summarized in simple stochastic transitions, and that including filter barriers are based on subjective decisions. However, significant progress has been made with the construction of this model, and although analysis of the stochastic process still leave many questions unanswered, the scaling seems appropriate for the purpose of modelling plant community assembly at the landscape level. With this model a landscape as a whole could be studied, and interconnections and feedbacks between two main processes, i.e. community assembly and disturbance, which would otherwise stay obscure, can become visible. More thorough knowledge on species dynamic processes at a stand scale, such as growth and dispersal of species, will uncover the underlying ecological processes of community assembly.

Community assembly determines what happens and where, given environmental and biotic constraints. Restoration defines a goal and a timeframe for management measures and at this point any threshold in the system becomes important. If we want to know the outcome of restoring communities then we need to have a model, which shows us certain constraint on community membership and reasons for changes. In other words, we need to be able to predict as far as is possible. The probabilistic Markov models of succession were criticized for being too rigid (Usher 1981; Hobbs 1983). In particular, the requirement that transition probabilities be specified as a constant value over time, conflicts with stochastic reality of year to year changes in the dynamics of natural systems. Anand and Orlóci (1997) suggested that the global pattern of state transition from determinism to apparent 'noise' could be explained, in the long term, by a specific combination of deterministic and random processes, i.e. the stationary Markov Chain plus quasi-random perturbation. Anand and Heil (2000) showed this approach in an analysis of a recovery process of heathland community. Petraitis and Latham (1999) emphasized the importance of scale and disturbance in the origin and persistence of alternative community assemblages. They suggested that this could occur in two ways – one through the 'natural' process of succession, and the other through a major disturbance event. The results nicely distinguish these two cases. In this Chapter a combination of deterministic and stochastic Markov model has shown to be able to predict fairly well changes in a real life example. That patterns and processes are scale dependent is certainly not new to the field of community ecology.

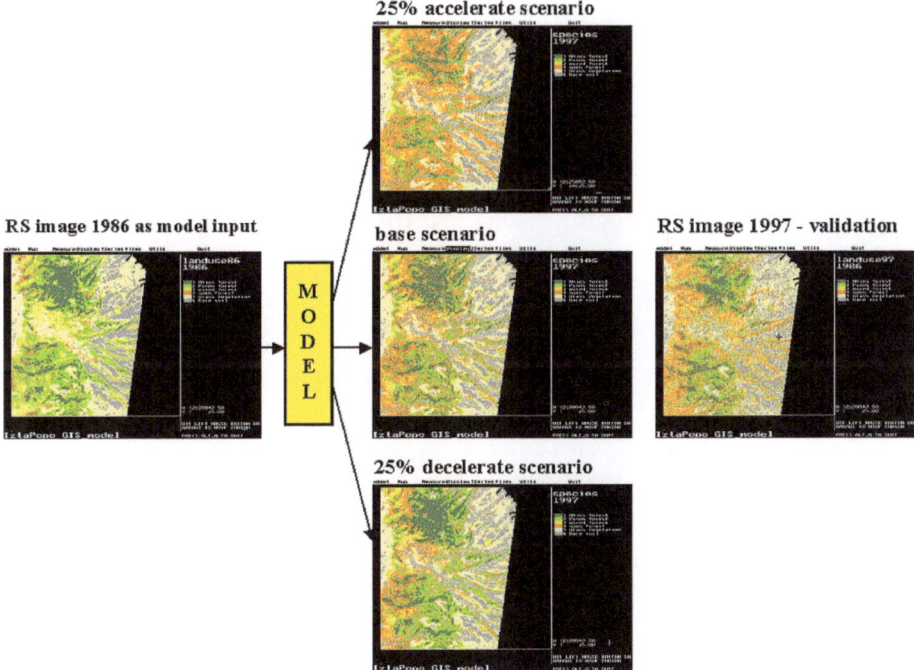

Figure 6.9. Examples of accelerated and decelerated deforestation scenarios for restoration purposes.

However, the nature of this scale dependency or 'scaling' is poorly understood. Here, we use empirical data to demonstrate that a process at finer spatial scale can help in understanding the dynamics at a higher spatial scale. For restoration purposes of the model we need to re-couple biotic and abiotic influences on community development and define stepwise goals that encapsulate critical points in the development process (see Figure 6.9). This can be done in three ways: passive restoration, interseeding or re-introducing of keystone species. Depending on the goals, possible outcomes can be simulated by adjusting the transition probabilities to restoration measures as mentioned above (Samuels and Lockwood, 2002). Future work will be directed towards the application of this model to analyse the potential impact of ecological restoration in the Iztac–Popo area.

REFERENCES

Anand, M. (1994). Pattern, process and mechanism – fundamentals of scientific inquiry applied to vegetation science. Coenoses, 9: 81-92.

Anand, M. and Orlóci, L. (1997). Chaotic dynamics in a multispecies community. Ecological and Environmental Statistics, 4, 337-344.

Anand, M. and Heil, G.W. (2000). Analysis of a recovery process: Dwingelose Heide revisted. Community Ecology, 1(1), 65-72.

Bart, J. (1995). Acceptance criteria for using individual-based models to make management decisions. Ecological Applications, 5, 411-420.

Belward, A.S. and Valenzuela, C.R. (1991). Remote Sensing and Geographical Information Systems for resource management in developing countries. Eurocourses Remote Sensing, vol. 1. Kluwer Academic Publishers, Dordrecht.

Burrough, P.A. (1986). Principles of Geographical Systems for Land Resources Assessment. Clarendon Press, Oxford.

Caswell, H. (1989). Matrix Population Models: Construction, Analysis, and Interpretation. Sinauer Associates, Inc. Publ. Sunderland, Masachusetts.

Chávez, J.M. and Trigo, N. (eds.) (1996). Programa de Manejo para el Parque Nacional Iztaccíhuatl-Popocatépetl. Colección Ecología y Planeación. Universidad Autónoma Mertopolitana, Xochimilco, Mexico.

Chesson, P. and Huntly, N. (1997). The roles of harsh and fluctuating conditions in the dynamics of ecological communities. American Naturalist, 150 (5), 520-534.

Clements, F.E. (1916). Plant Succession. An analysis of the Development of Vegetation. Carnegie Intitute Publication Number 242 Washington, D.C.

DeAngelis, D.L. and Gross, L.J. (1992). Individual-based Models and Approaches in Ecology Populations, Communities and Ecosystems. Chapman & Hall. London.

Diaz, S., Cabido, M. and Casanoves, F. (1998). Plant functional traits and environmental filters at a regional scale. Journal of Vegetation Science, 9(1), 113-122.

Diaz, S., Cabido, M., Zak, M., Martínez Carretero, E. and Araníbar, J. (1999). Plant functional traits, ecosystem structure and landuse history along climatic gradients in central-western Argentina. Journal of Vegetation Science, 10, 651-660.

Eastman, J.R. (1993). Idrisi manual. Clark University, Graduate School of Geography, Worcester, USA.

García, M., Lopez-Paniagua, J. and Rosas, M. (1992). Socioeconómicos. In: SEDESOL-Banco Mundial-UAM-X. Programa de manejo para el Parque Nacional Ixtaccíhuatl-Popocatépetl.

Grime, J.P. (1979). Plant Strategies and Vegetation. Chichester, John Wiley.

Heil, G.W. and Van Deursen, W.P.A. (1996). Searching for patterns and processes - modelling of vegetation dynamics with Geographical Information Systems and Remote Sensing. Acta Botanica Neerlandica, 45, 543-556.

Hobbs, R.J. (1983), Markov models in the study of post-fire succession in heathland communities. Vegetatio, 56, 17-30.

Hobbs, R.J. (1994). Dynamics of vegetation mosaics: can we predict responses to global change? Ecoscience, 1, 346 – 356.

Hobbs, R.J. (1997). Can we use plant functional types to describe and predict responses to environmental change? In: Smith, Shugart, Woodward (eds). Plant Functional Types. pp 66-91. Cambridge University Press.

Hobbs, R.J. (1998). Restoration of disturbed ecosystems. In: Ecosystems of the World 16. Disturbed Ecosystems, (ed) Walker, L, pp 673-87. Elsevier, Amsterdam.

Hogeweg, P. and Hesper, B. (1990). Individual-oriented modelling in ecology. Mathematics and Computer Modelling, 13, 83-90.

Huisman, J. and Weissing, F.J. (1999). Biodiversity of plankton by species oscillations and chaos. Nature, 402, 407-410.

Huston, M., DeAngelis, D.L. and Post, W. (1988). New computer models unify ecological theory, BioScience, 38, 682-691.

Jensen, J.R. (1986). Introductory Digital Image Processing. Prentince-Hall, Englewood Cliffs, New Jersey, USA.

Jörgensen, S.E. (1988). Fundamentals of Ecological Modelling. Developments in Environmental Modelling, Vol.9. Elsevier, Amsterdam.

Law, R. (1999). Theoretical aspects of community assembly. In: J. McGlade. (ed). Advanced Ecological Theory: principles and applications. pp. 143 – 171. Blackwell Science.

Lilesand, T.M. and Kiefer, R.W. (1994). Remote Sensing and Image Interpretation. John Wiley & Sons, Inc. 750 p.

Matson, P.A.and Ustin, S.L. (1991). The future of remote sensing in ecological studies. *Ecology*, 72, 1917 -1945.

Noble, I.R. and Slayter, R.O. (1980). The use of vital attributes to predict successional changes in plant communities subject to recurrent disturbances. Vegetatio, 43, 5-21.

Orlóci, L., Anand, M. and He, X.S. (1993). Markov chain: a realistic model for temporal coenosere? Biométrie-Praximétrie, 33, 7-26.

Petraitis, P.S. and Latham, R.E. (1999). The importance of scale in testing the origins of alternative community states. Ecology, 80, 429-442.

Ridgley, M. and Heil, G.W. (1998). Multicriterion Planning of Protected-Area Buffer Zone: An Application to Mexico's Izta-Popo National Park. In: E. Beinat and P. Nijkamp (eds). Multicriteria evaluation in land-use management. pp 293-311. Kluwer academic publishers, Dordrecht.

Rosenzweig, M.L. (1995). Species diversity in space and time. Cambridge University Press.

Samuels, C.L. and Lockwood, J.L. (2002). Weeding out surprises: Incorporating uncertainty into restoration models. Ecological Restoration, 20(4), 262-269.

SEDESOL-Banco Mundial-UAM-X (1992) Programa de manejo para el parque nacional Iztaccihuatl-Popocatepetl. Mexico City: Secretaria de Desarrollo Social.

Trigo, N., Bobbink, R. and Heil, G.W. (1995). A Unified Framework for Ecological Monitoring of the Izta-Popo National Park, Mexico. In: T.B. Herman, S. Bondrup-Nielsen, J.H.M. Willison and N.W.P. Munro (eds). Ecosystem Monitoring and Protected Areas. pp 575-580, Science and Management of Protected Areas Association, Wolfville, Nova Scotia, Canada.

Usher, M.B. (1981). Modelling ecological succession, with particular reference to Markovian models. Vegetatio, 46, 11-18.

Van der Maarel, E. and Sykes, F.S. (1993). Small-scale plant species turnover in a limestone grassland: the carousel model and some comments on the niche concept. Journal Vegetation Science, 4, 179-1988.

Van Deursen, W.P.A. (1995). Geographical Information Systems and Dynamic Models. PhD. Thesis, University of Utrecht, department of physical geography. 198 p.

Van Deursen, W.P.A. and Heil, G.W. (1993). Analysis of heathland dynamics using a spatial distributed GIS model. Scripta Geobotanica, 21, 17-27.

Van Hulst, R. (1979). On the dynamics of vegetation: Markov chains as models of succession. Vegetatio, 40, 3-14.

Velazques, A., Romero, F.J, Rangel-Cordero, H. and Heil, G.W. (2001). Effects of landscape changes on mammalian assemblages at Izta-Popo Volcanoes, Mexico. Biodiversity and Conservation, 10, 1059 – 1075.

White, P.S. and Pickett, S.T.A. (1985). Natural disturbance and patch dynamics, and introduction. In: S.T.A Pickett and P.S. White(eds). The Ecology of Natural Disturbance and Patch Dynamics. pp 3-13. Academic Press, New York.

White, P.S. and Jentsch, A. (2001). The search for generality in studies of disturbance and ecosystem dynamics. Annals of Botany, 62, 399-450.

Wilson, J.B. (1999). Assembly rules in plant communities. In: E. Weitherand P.Kedd (eds). Ecological Assembly: Advances, Perspectives Retreats. pp 130-164. Cambridge University Press.

Chapter 7

Remote Sensing biomass of forested ecosystems: Modelling the carbon cycle of the Iztaccíhuatl - Popocatépetl National Park, México

Gerrit W. Heil*, Roland Bobbink**, Nuri Trigo Boix*** and Betty Verduyn*

*Department of Plant Ecology, **Department of Landscape Ecology, Faculty of Biology, Utrecht University, The Netherlands, ***Departamento El Hombre y su Ambiente, UAM-X, México

Key words: remote sensing, carbon cycling, modelling, Climate Change

Abstract: Increases in the concentration of carbon dioxide, as well as of other so-called greenhouse gases, in the Earth's atmosphere will lead to climate change. So far, policy responses to limit the effects of climate change have focused on the reduction of emissions of greenhouse gases at source. Another possibility would be to enhance the capacity of ecosystems to act as carbon sinks, such as by increased afforestation, in order to reduce carbon dioxide (CO_2) concentrations in the atmosphere. Many countries, especially in the Third World, lack sufficient information on the actual ground cover of different types of (agro) ecosystems, i.e. on carbon pools in the form of terrestrial ecosystems. Without such land cover data, it is almost impossible to analyze properly the non-fossil cycles of these countries. However, such data can be obtained with remotely sensed data from satellite images. To obtain insight into the effects of changes in land cover and the time-scales over which they occur, a dynamic simulation model on carbon cycling of terrestrial ecosystems has been converted into a GIS environment. The results of our case study show a significant correlation between field data on the amount of biomass/carbon of different natural forests and NDVI values from a satellite image of the study area. The results on field data and on NDVI data in combination with production and decomposition rates from literature have been used to calibrate the simulation model. After a sensitivity test, the model has been applied to study the effects of fragmentation on the carbon cycle of the Iztaccíhuatl - Popocatépetl National Park and its surrounding buffer areas.

G.W. Heil et al. (eds.), Ecology and Man in Mexico's Central Volcanoes Area, 147–171.

1.   INTRODUCTION

It is now widely recognised that carbon dioxide and several other atmospheric trace gases such as methane, chlorofluorocarbons, and nitrous oxide, play an important role in determining the temperature of the Earth (Tegart *et al.* 1990). In contrast with both nitrogen and oxygen, which together make up more than 99% of the atmosphere, these trace gases absorb infrared radiation, or radiant heat. Because the total amount of so-called greenhouse gases (GHGs) is small, their concentrations are easily changed. There is strong evidence that there is a direct correlation between increased emissions of CO_2, methane and other GHGs, and the rise in global temperature (Bolin 1986). This phenomenon has been called global warming by the greenhouse effect (e.g. Houghton and Woodwell 1989; Lashof and Tirpak 1989; IPCC 1997; Bonan 2002).

The increase in CO_2 is the outcome of a series of interactions between the atmosphere, the oceans, the terrestrial biota, human activities, and geological activity (IPCC 1997). The two most important sources of CO_2 are the combustion of fossil fuels and the decay (or combustion) of biotic residues on land. Some of the carbon from these two sources accumulates in the atmosphere; some is transferred to terrestrial biota and the oceans. At present, however, the rate of carbon transfer to terrestrial biota and the oceans is lower than the rate of release of CO_2 into the atmosphere, with the result that atmospheric concentrations are increasing.

This increment, together with the accumulation of other greenhouse gases, could produce that the mean temperature of the Earth rises from 1.5 to 4.5°C by this century, a temperature increase without precedents in the history of the Earth (IPCC 1997).

The most important biotic source of an increase in CO_2 is the destruction of forests. A review of the biotic contribution suggests a net release to the atmosphere of $2\text{-}5\times10^{15}$ g C annually from deforestation (Watson *et al.* 2000). Most of the carbon stored in the earth's biota and organic soil layers is associated with forests so that a small change in the amount of carbon stored on these ecosystems might change the atmospheric concentration significantly. The extent to which the biota and the organic soil layers of earth as a whole are changing is obviously important in trying to predict the future CO_2 content of the atmosphere (Woodwell 1984; Caspersen *et al.* 2000; Chomitz 2000).

Although there are many uncertainties about the consequences of increased concentrations of GHGs, the risk of potentially damaging effects has induced governments to adopt a precautionary approach and to formulate policies to reduce emissions. The Kyoto Protocol aims to reduce net emissions of GHG's to the atmosphere by various measures including land-use management (UNFCCC, 1997). Since CO_2 is the most important contributor to greenhouse gas emissions at the global level, discussions have focused strongly on reducing emissions of CO_2 from fossil fuel combustion.

Efforts to limit the effects of climate change have strongly focused on the reduction of emissions of CO_2 and other greenhouse gases into the atmosphere. Another possibility is to enhance the capacity of ecosystems to act as carbon sinks such as by increased afforestation. Although carbon sink enhancement might be a relevant option to reduce CO_2 concentrations, the potential contribution of afforestation schemes to this reduction is ambiguous (Kirschbaum *et al.* 2001). In contrast, still large amounts of carbon are lost because of land use changes such as deforestation. Changes in the non-fossil carbon cycle are strongly delayed because of different feedback mechanisms, such as decomposition and mineralization. Although evaluation of land use changes uses an equilibrium approach, the time scales to reach such equilibrium are of the order of tens of years. Consequently, effects of land use changes should be evaluated as a function of time, through a dynamic approach. Among a range of forest carbon models (e.g. Mohren and Klein Goldewijk 1990; Kirschbaum 1999)(see e.g. reviews by Bataglia and Sands 1998; Kirschbaum and Mueller 2001), the simulation model RANK has been developed to evaluate the effect of land use changes (Heil and Gerrits 1992; Heil *et al.* 1993). RANK is a dynamic model that is based on ecosystem growth characteristics, and captures the transient responses of ecosystems to changes in land use.

To evaluate the effect of carbon losses and gains, data on changes in land-use and land cover are crucial because, depending on the way the land is used in a certain area, it may act either as a sink or a source of CO_2. Many countries, particularly in the Third World, lack sufficient data on land use, i.e. data on the (historical and/or actual) ground cover of different types of (agro) ecosystems (Walker and Steffen 1996). Without such land-use data, it is almost impossible to analyse adequately the non-fossil carbon cycle of a particular area. However, such data can be obtained using remote sensing (RS) techniques.

RS techniques have been used to facilitate in land use surveys since the early 1970s (Weismiller and Kaminsky 1978). They are based on the strong relationship between the spectral reflectance characteristics of the Earth's surface and many of its physio-chemical properties. Estimates of biomass are made based on the reflectance characteristics of the vegetation canopy. Several studies have demonstrated that the determination of biomass with RS techniques is, in fact, a relatively simple procedure (Aldred 1976; Erb 1980; Strahler 1981; Tucker and Miller 1979; Woodwell 1984; Pinty and Verstraete 1992).

In this study a methodology has been developed for the application of RS techniques in combination with a geographical information system (GIS) to obtain important data on non-fossil carbon cycles, i.e. actual data on biomass, in data-poor countries.

A spatial model of the non-fossil carbon cycle - RANGIS - is developed and implemented in a GIS environment. RANGIS is able to calculate changes in carbon fluxes and stocks in combination with RS information on land cover and initial biomass values. In this study, RS information on the Iztaccíhuatl - Popocatépetl area has been used to develop the methodology for the implementation of land use/ land cover in the carbon model with distinctly different types of ecosystems. In this way, the potential effect of carbon release as a result of deforestation and forest fragmentation has been analysed.

2.   DESCRIPTION OF THE STUDY AREA

The Iztaccíhuatl and Popocatépetl volcanoes are located in the central part of the Transversal Neovolcanic Axis in Central Mexico, between the geographic co-ordinates of 18° 59' and 19° 16' 25" N and 98° 34' 54" and 98° 42' 08" W. The mountains in the study area range from 2,500 to 5,452 meters, and the surface area of the study area covers approximately 1,580 km2 (Figure 7.1).

The parental material includes volcanic ashes, tubas and pumic stone. The soil is mainly of the andosol type, with vitric, humic, molic and ocric horizons. The water resources of the zone come mainly from the snow thaw and rainfall, which is quite high in summer. This allows the formation of a few permanent streams and several temporal ones.

According to meteorological stations near the mountains, they have three different climates: the humid or sub-humid temperate with precipitation during the summer; the semi-cold with precipitation during the summer; and the cold above 4,000 m (see also Chapter 1).

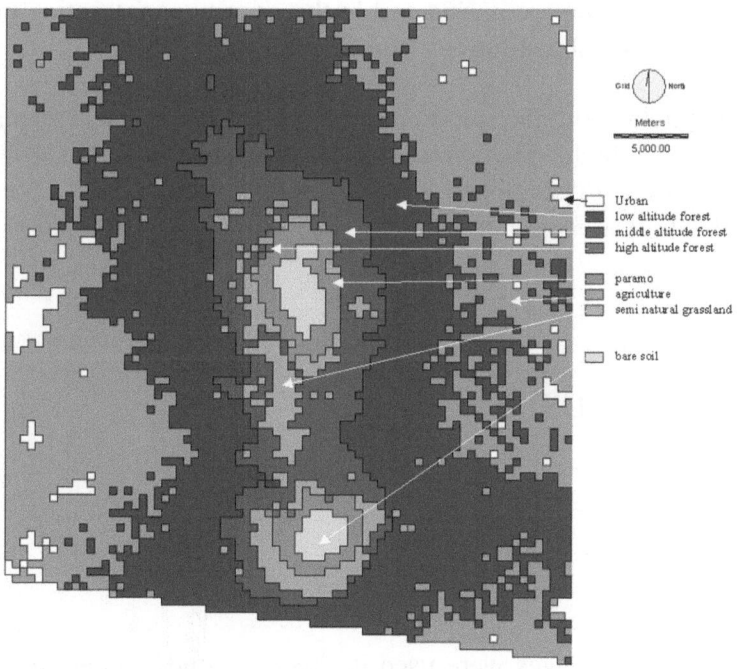

Figure 7.1. Map of the land cover of the Izta-Popo area (1 pixel is 500 x 500 m).

The Iztaccíhuatl and Popocatépetl volcanoes are very important from the phyto-geographical and botanical perspectives (see Chapters 2, 3, and 4). That is due to their location in the Transversal Neovolcanic Axis in

Central México, which includes the highest mountains of the country and has a marked altitudinal gradient. With these characteristics, several vegetation types with an important floristic richness are present in the area. These vegetation types are: coniferous and oak forests which are the most rich biomes considering their number of species, alpine vegetation with a high abundance of endemic species, and mountainous mesophile forest within the range of 2,500 to 2,800 m, which is found only in discontinuous patches and under very specific environmental conditions.

From the zoogeographic point of view, the volcanoes belong to the contact and transitional area between the neartic and neotropical wildlife. Therefore, there is a high richness of species in the region, which represents up to 30% of the total of mammals in the country (Volcano-Transverse Biotic Province, see Chapter 5). About the birds, there are many resident species, some of which are considered in the priority conservation projects of several international institutions (see Chapter 4).

As a final note about the importance of the volcanoes, their recreational potential is very high. That is one of the reasons why, in 1935, the Iztaccíhuatl and the Popocatépetl were declared a National Park. The actual limit of the Park is at 3,600 m.

3.   METHODS

3.1   Using RS data

Remote sensing images were used in this case study to derive two sets of information needed for the Non-Fossil Carbon Modelling. The first type of information derived from the images is an ecosystem classification. The second type of information is a vegetation index, which is used as an estimator of initial amounts of carbon in the different compartments.

The image used is a MSS (Multi Spectral Sensor) image from January 1986 with a ground resolution of 80 x 80 m. Images recorded just before the rainy season starts seemed to be favourable, since the contrast between natural and agricultural vegetation would be at an optimum. The image is georeferenced using 1:50.000 topographic maps of the volcanoes area (INEGI 1978, 1982, 1983 and 1985a, b and c) and after ecosystem classification resampled to a 500 x 500 meter resolution by a majority algorithm.

The georeferenced MSS image was classified for ecosystems using a supervised classification scheme. For the model, four ecosystem classes were derived from the image (urban area, forest, grasslands, and agricultural area/bare soils).

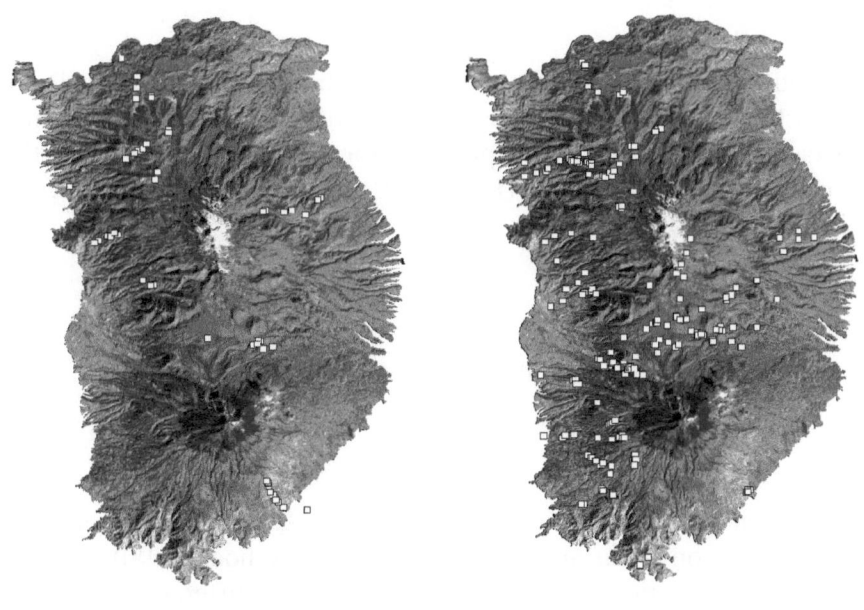

Figure 7.2. Validation (left) and sampling (right) points in the Iztac - Popo area

Because of the acquisition date of the image, no distinction could be made between the mountainous bare soils above 4,000 meter and the agricultural area, which was at the date of acquisition just starting to grow crops. A further classification based on the combination of these intermediate results and the digitised elevation map yielded the following ecosystem classes:

- urban area
- low altitude mountainous forest (LA forest, < 3,200 m)
- medium altitude mountainous forest (MA forest, 3,200–3,600 m)
- high altitude mountainous forest (HA forest, 3,600–3,900 m)
- semi-natural grasslands (grazed grasslands below 3,600–3,900 m)
- paramo (natural grasslands > 3,600 m)
- agricultural area
- bare soils

Ground truth for the ecosystem classification was gathered using the results of the fieldwork campaigns, interpretation of the topographic maps, and data from the Management program for the National Park Izta-Popo (Chávez and Trigo, 1996) (Figure 7.2). The resulting ecosystems classification is shown in Figure 7.1.

3.2   Deriving NDVI-maps for the assessment of biomass/carbon

Vegetation indices are very useful for the vegetation monitoring and biomass estimations (Curran, 1980, Clevers and Verhoef, 1990). The most widely used vegetation index is the Normalized Difference Vegetation Index (NDVI), which is derived for each individual pixel in the image using the formula:

NDVI = (MSS4-MSS3) / (MSS4+MSS3)

MSS3 = value of MSS Band 3
MSS4 = value of MSS Band 4

The georeferenced bands are used to derive the NDVI for the area. The NDVI image for the study area is shown in Figure 7.3.

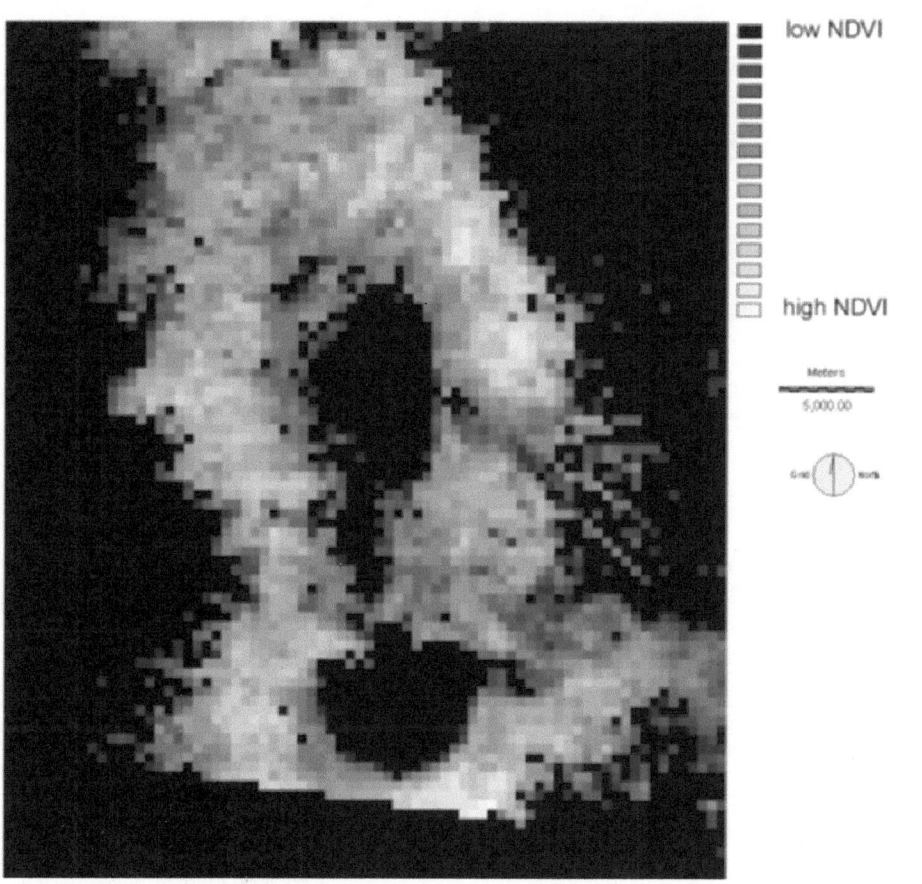

Figure 7.3. NDVI image of the study area

3.3   Field work

Figure 7.4 shows the general procedure that was followed to obtain carbon content values for each compartment per system. This includes field sampling, laboratory analysis and data processing.

The NDVI image of the Izta-Popo area has been used for the selection of the sampling plots and number. This image has been used to define systems with significantly different vegetation density, i.e. with a strong difference in NDVI-values. In order to obtain plots that were representative for the landscape as a whole, cover and uniformity of the vegetation were the criteria for the plot selection.

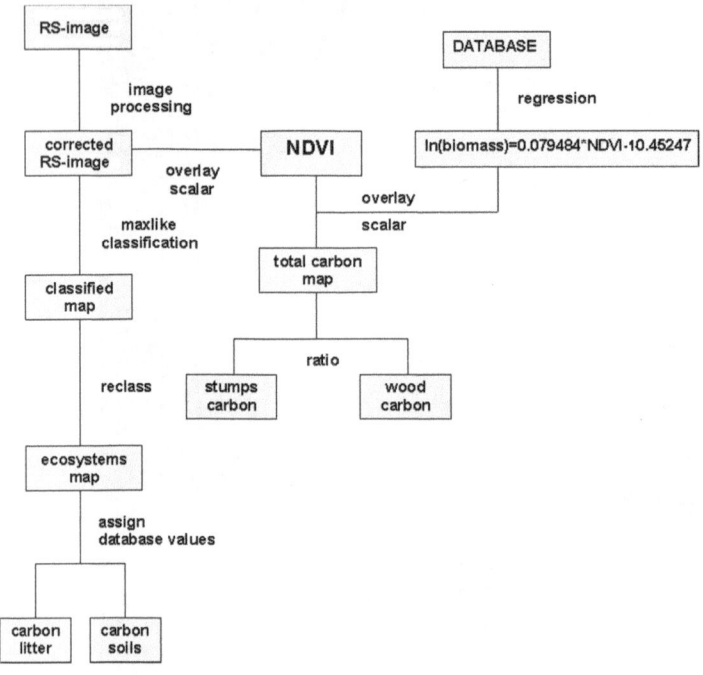

Figure 7.4. Diagram of the methodology

Measurements for biomass and carbon content have been carried out in January 1993. Bulk samples consisting of five sub-samples were taken from the different natural vegetation types (7 or 8 bulk samples of each type). With a Global Positioning System (GPS) the geographical position has been determined as close as to 10 -20 m from each sampling plot. In each plot, all the trees with at least 10 cm in diameter at breast height (DBH) and more than 5 meters high were measured. For the understory and the herb layer, litter and biomass were collected in 4 or 5 random sub plots in quadrants of 50 X 50 cm. The litter was collected by hand and the biomass by harvesting.

Soil samples were collected with a soil auger in the different forests and grasslands types. The samples were obtained as deep as the soil organic layer. Therefore, the depth of these samples was variable depending of the slope, the soil and vegetation types.

Since the agricultural areas were not used in this period because of the time of the year, only soil samples were taken in the agricultural systems.

3.4   Laboratory work

The aim of the laboratory work was to obtain the values of the content of carbon in the different compartments measured in the field. The final results are expressed in C tons/ha. For that purpose, the compartments were divided in:

o   soil
o   litter
o   herb layer
o   trees

The determination of soil organic carbon content is obtained from the loss-on-ignition (LOI) values as a measure of organic matter. The conversion factor of 58% in soil organic matter has been taken from Moore and Chapman (1986).

For the calculation of the carbon content of the litter and the herb compartments, the samples were air dried for 72 hours, weighed, and the carbon content was calculated according to standard conversion factors, i.e. 0.48 and 0.5 for biomass and litter, respectively (Moore and Chapman 1986; Chambers *et al.* 2000; Hughes *et al.* 1999)

The calculations of carbon content in the tree layer were based on regression equations for estimating total above ground biomass based on DBH and height. To obtain the carbon content of each plot, a conversion factor for carbon in trees was used, i.e. the conversion factor of 0.48 of the dry biomass. Finally, the results are expressed in ton C/ha and the data on carbon content of soil, litter and above ground biomass were used for a database, which was arranged by vegetation type.

4.   SHORT DESCRIPTION OF THE CARBON CYCLE MODEL

In order to evaluate the potential effectiveness of carbon sink enhancement measures, a carbon cycle model has been constructed (Heil *et al.* 1993). This model calculates carbon fluxes such as production, decomposition and fluxes into the soil compartment. The starting conditions in 1992, calculated from historical land-use statistics, supply the model with the initial carbon stored in four "stock compartments" (the soil, wood, stumps and litter) of the various ecosystems. On the basis of the initial carbon storage data, fluxes to and from these compartments can be calculated to determine the amounts of carbon stored at any particular moment. The amounts of carbon stored and released by decomposition are the data outputs of the model. The model consists of simple first order equations between the different stocks (Heil and Gerrits 1992).

The reliable quantification of carbon fluxes in the various ecosystems is strongly dependent on the initial amounts of carbon stored in the stock compartments (*c.f.* Caspersen *et al.* 2000). Provided that the system is in equilibrium, it is possible to calculate the amount of carbon stored in the soil compartment when the flux into the soil is known, because the efflux depends on the carbon storage capacity. However, the assumption that ecosystems are in equilibrium is not realistic.   To estimate the carbon stored in the various systems in 1992, it is therefore necessary to calculate the carbon fluxes and stocks from past data. With the carbon model, carbon stocks are calculated from 1900 up to 1992. The calculated carbon stocks of 1992 are used as initial stocks of the model.

In order to be able to use the carbon cycle model in a spatial environment, a spatial model has been developed on basis of the original model. The database of the spatial model consists of different raster GIS maps with a resolution depending on the total surface of the area. The total area covered by the GIS database depends on the format of the map used. The first map is a digital land use – land cover map created and updated entirely with remote sensing data (Figure 7.1). For our case study, the resulting raster map was resampled to a 500 x 500 m grid using a majority algorithm. The number of land use types in the model can be adjusted to correspond with the land use types that have been classified with the RS images.

Other maps, such as a soil map or an altitude map, might be included to calculate for example production and decomposition rates. However, in our case study values for production and decomposition rate for the different ecosystems are taken from literature (Lieth and Whittaker 1975).

Another map is the database link map, a raster map that contains pointers to the records in the carbon stock and fluxes database. This map is responsible for the link between data in the GIS and in the database.

The carbon stock and fluxes database is used to store the values of the accumulated stocks of carbon in the compartments of each ecosystem. For each time step both the database and the flux fields for each of the records are updated. The database is linked to the GIS maps through the database link map, which contains pointers to each of the records in the database. The database contains the following fields for the storage of carbon within each ecosystem:

· Wood: the amount of carbon stored in the wood compartment for each individual grid cell (ton C/ha);

· Stumps: the amount of carbon stored in the stumps compartment for each grid cell (ton C/ha);

· Litter: the amount of carbon stored in the litter compartment for each grid cell (ton C/ha); and

· Soil: the amount of carbon stored in the soil compartment for each grid cell (ton C/ha).

An aggregated time series database is used to record the total amounts (the sum of all grid cells) of carbon stored in, and released from the various compartments. Each time step of the model adds one record to this database, so that at the end of the run the database contains a time series of stocks and fluxes.

5.   THE EFFECT OF FRAGMENTATION ON THE CARBON STATUS

5.1   Relation between NDVI and biomass/carbon

The field data on biomass/carbon of the different sampling plots in the Izta-Popo area have been used for comparison with the NDVI data of the RS image from the same locations as the sampling plots. For the analysis, the data are ln-transformed, because one can expect an S-shape relationship between NDVI and biomass/carbon as a result of saturation in reflection when biomass/carbon values are high. The ln NDVI-values from the RS image are significantly related with the measured amounts of aboveground biomass/carbon (R^2=0.74, df = 20). Using this correlation yields the following equation to calculate the above ground biomass/carbon from the NDVI-values:

$$\ln \text{ ton C/ha} = 0.079484 * \text{NDVI} - 10.45247$$

Where: ln = natural log; ton = 1000 kg; C = carbon; ha = hectare; and NDVI = Normalized Difference Vegetation Index value.

Depending on the type of ecosystem, this above ground biomass has been distributed over the compartments wood, stumps and litter. Initial soil stocks were assigned as average values for the individual ecosystems, because no spatial distinction can be made on basis of the RS-image (Table 7.1).

A different approach to verify the applicability of NDVI was done with a series of fisheye photos (Figure 7.5). For this fisheye photos were taken from sample plots to compare with NDVI values. From each plot, also the GPS position was taken during a ground truthing of the supervised classification in 1998.

The photos were quantified with the analysing software Winphot (Ter Steege, 1994) and then the leaf area index (LAI) was calculated. The LAI of the fisheye photos was tested against the NDVI-values from the RS image of the same plots. The results show a significant relationship between LAI and NDVI (R^2 = 0.34 df = 32). The LAI values are used as an independent estimator for the relation between NDVI-values and LAI/biomass of the vegetation (Figure 7.6).

Once the model had been calibrated, the next steps have been carried out: i.e. a sensitivity analysis of the model and an analysis of the carbon cycle as a result of changes in land use by deforestation/fragmentation

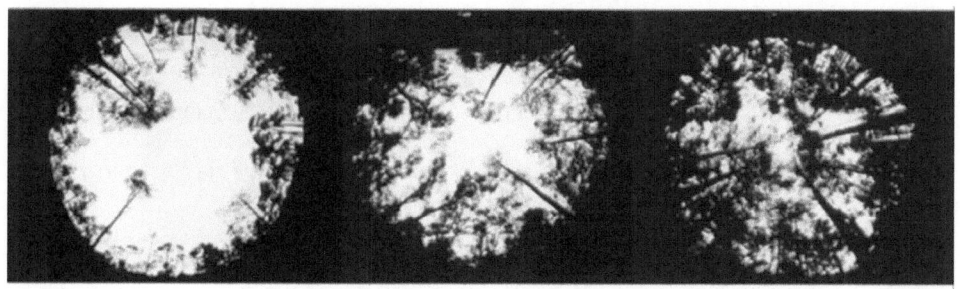

Figure 7.5. Fish eye photo's of forest canopies

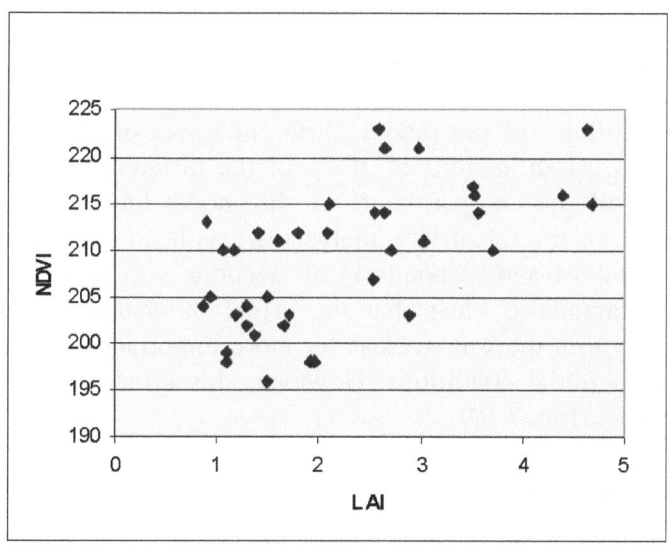

Figure 7.6. Relation between LAI from fisheye photo's and NDVI-values

Table 7.1. Conversion equations of NDVI-values to carbon data of the different compartments as used for the case study.

	Carbon (1) (tons/ha)	wood (tons/ha)	stumps (tons/ha)	litter (tons/ha)	soil (tons/ha)
Urban		0	0	0	0
LA forest	f(NDVI)	0.70 * (1)	0.25 * (1)	0.05 * (1)	196
MA forest	f(NDVI)	0.70 * (1)	0.25 * (1)	0.05 *(1)	167
HA forest	f(NDVI)	0.70 * (1)	0.25 * (1)	0.05 * (1)	191
Paramour	f(NDVI)			1.0 * (1)	20.7
Agriculture					109
Grassland	f(NDVI)			1.0 * (1)	270
Bare Soils		0	0	0	0

Where f = function to convert NDVI values to carbon in tons/ha

5.2   Sensitivity Analysis

To test the sensitivity of the model, different series of simulations have been carried out. Modifications of 30 % of the different rates and of the initial amounts of the compartments of the model have been executed. From the results of the sensitivity analysis, it has been concluded that the model is very robust and responds in an accurate way to the introduced changes. It became also clear that the effect of changes in the initial amounts of carbon in the soil stock is far more important than variation in any of the other initial conditions. However, this effect was levelled off after 10 - 15 years (Fig. 7.10).

5.3   Fragmentation

Effects of land use changes to carbon emission can only be evaluated if they can be compared to a base case, which means that the land use remains the same during the simulation periods. For our study, it was interesting to compare this base case with the fragmentation as described in Chapter 3, which should affect the carbon balance in the system. A satellite image of 1997 on the fragmentation of the research area has been resampled to the same resolution of the base case (Figure 7.7).

This resulted in our case study that the deforestation towards agriculture, grazed grasslands and urban areas yields approximately 9% (i.e. 3,600 ha, 2,500 ha and 150 ha, respectively) of the original forest in 1986. Another change of land use was caused by the change of the agriculture to urban areas (1,250 ha) (see also Chapter 3).

6.   RESULTS

The spatial distribution of the different land use types under the base case and under fragmentation is represented on the land use map of Figure 7.1 and Figure 7.8, respectively. These maps were used as input for the model, which has been run for the period 1992 to 2015. To analyze the effects of fragmentation the results for each compartment are separated, i.e. carbon in wood, in stumps, in litter, and in soil.

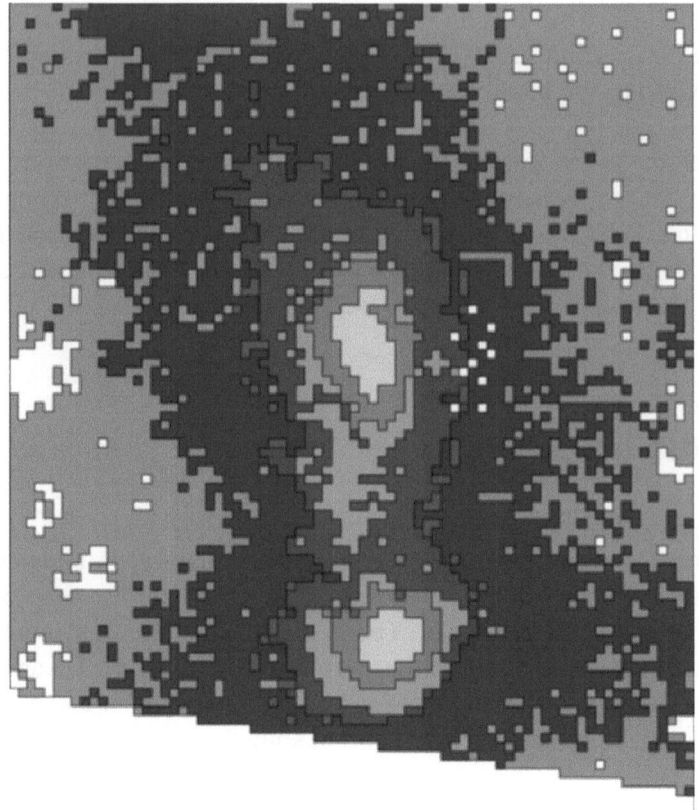

Figure 7.7. Fragmentation scenario. For legend see Figure 7.1.

6.1   Results of the base case

The map with the spatial results shows that the model responds well and that the carbon stocks in wood of the different pixels have different biomass due to the different initial values and different growth ratios (Figure 7.8). Considering each compartment in a separate analysis, the results show that a significant change of carbon in all compartments occurred. The results also show that the simulation period of 23 years is too short to come at a stabilization point for most of the compartments. For the stumps, there is a strong decrease of carbon content during the first eight years. This process is more or less similar for the litter compartment. This is due to the time scale in which the carbon changes take place in these compartments. In addition, it should be realised that these two compartments have potentially the smallest carbon stock of forest ecosystems, and their proportional responses to changes from in- and output fluxes will be strongest. In the potentially largest carbon pool, i.e. the soil, there is a constant decrease under the base case circumstances. This can be explained by the deforestation of recent years/decades in the area, i.e. the carbon content is delaying behind the actual situation of the land use.

It is obvious that in some parts of the forests the above ground biomass, of which the initial values are estimated on the NDVI-values from the satellite image, are increasing in amount of carbon in wood because of the relatively young age of the forest at the beginning of the simulation period.

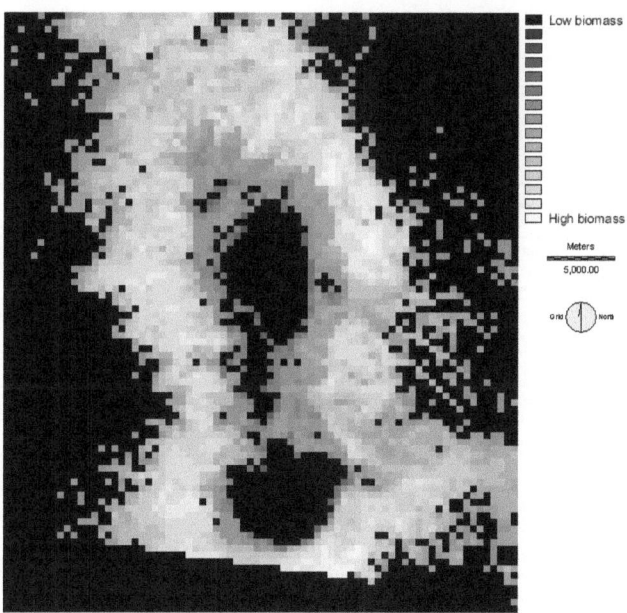

Figure 7.8. Calculated biomass for the study area

When equilibrium has been established, in-/decrease of carbon will not occur any more. In an equilibrium situation, there will also be no change of the forest soil carbon content anymore.

6.2   Results of fragmentation

The contrast between the results of the base case and the fragmentation simulation is striking, i.e. because of the fragmentation there are significant differences in carbon content between the two types of simulations. The most significant reduction in amount of carbon is in the wood and stumps compartments. Of course, for both stumps and wood the values change suddenly at the moment of fragmentation introduction in 1995. After this, the values of the stumps between the base case and fragmentation stay more or less equal over the time, but the difference in wood values increases slightly over the time.

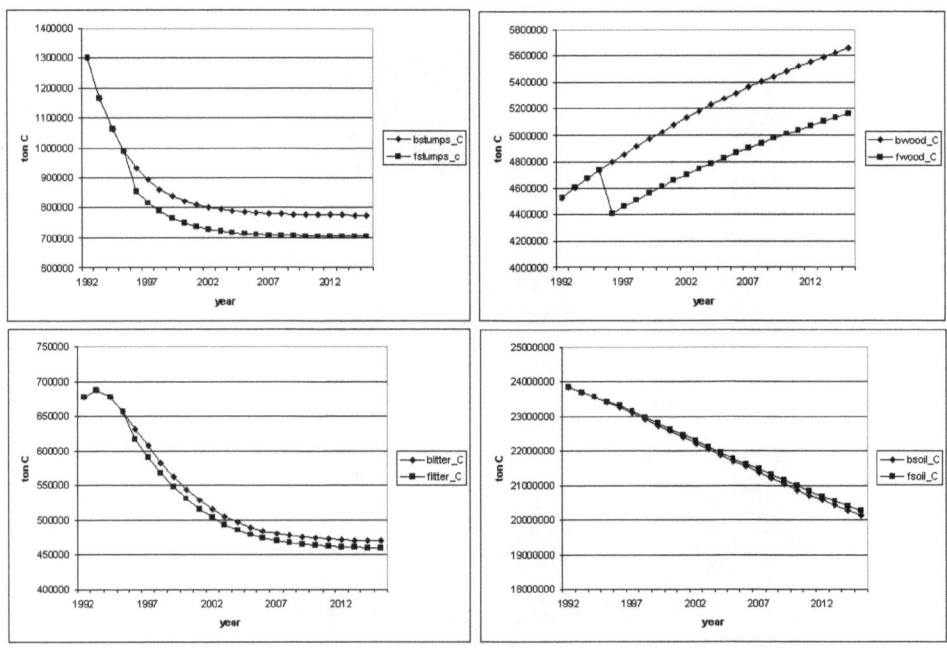

Figure 7.9. Results of the simulations (b- = base case; f- = fragmentation scenario). Results are from top left = stumps; top right = wood, bottom left = litter, and bottom right = soil.

For litter, the decrease of the carbon stock is not very significant and stabilizes rather quickly. The results show that the carbon stocks of the soil do not show big differences between the simulations.

To summarize the results as described above, the net carbon sink has been calculated under the two scenarios (Figure 7.9). As expected, there is a negative effect on the carbon content in the different compartments due to the fragmentation. This effect is not constant but changes over the time, i.e. the difference increases in the first 15 years and decreases later on (Figure 7.10). This is the result of the ingrowth effect in the open forests.

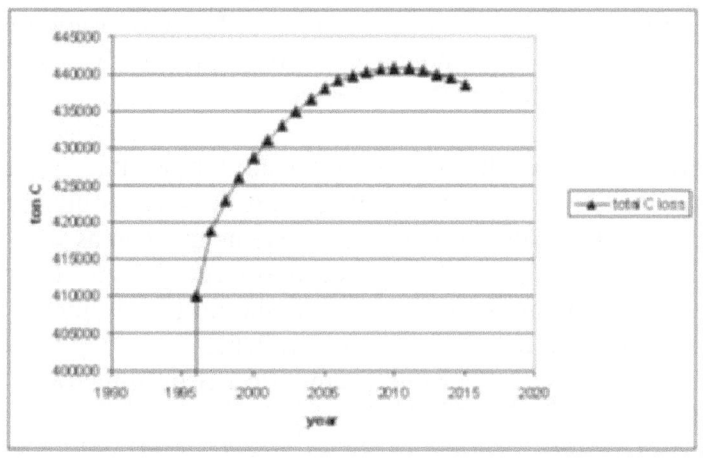

Figure 7.10. Total difference in C-loss between base case and fragmentation scenario.

7.   CONCLUSIONS

It should be recognised that the methodology developed is not scale dependent. It can be applied within the framework of regional planning, as shown in this case study, as well as in local and national planning. In the case of specific plans, more attention should be paid to the calibration of detailed RS information, because filling the GIS database with average initial values and production data will yield less specific results. In relation to the latter, the results of this case study showed that data on actual biomass/carbon could be obtained through RS images. It appeared that there is a significant relation between vegetation reflection characteristics, expressed in NDVI, and amount of above ground biomass/carbon. This relation has been expressed in a regression equation to be able to determine amount of biomass/carbon from NDVI values.

Results of a sensitivity analysis show that the RANGIS model is robust, when calibrated for this case study. To show the potential of the methodology developed, also a fragmentation scenario has been used to simulate effects of changes in land use on the carbon cycle of the Iztac-Popo volcanoes area. The results of the fragmentation scenario showed that –as could be expected - changes in land use by deforestation will have a negative effect on the amount of carbon stored in the different compartments and that there will be a net carbon release from the area.

In conclusion, the results of the case study show, that based on RS satellite data, the developed methodology can be successfully applied to analyse the non-fossil cycle of natural ecosystems from areas with strongly limited data on actual ground cover of these ecosystems. However, to improve the methodology for larger areas it should be considered to make use of multi temporal RS data e.g. as available from NOAA/AVHRR. Further research into the use of multi-temporal RS information is strongly recommended.

ACKNOWLEDGEMENT

We are grateful to Willem van Deursen (Carthago) for his contribution to the implementation of the carbon cycle model in the PCRaster GIS-environment.

REFERENCES

Aldred A.H. (1976). Measurement of tropical trees in large-scale aerial photographs, Forest Management Institute, Ottawa, Ontario, Canada.

Battaglia, M. and Sands, P.J. (1998). Process-based forest productivity models and their application in forest management. Forest Ecological Management 102: 13 – 32.

Bonan, G.B. (2002). Ecological climatology: Concepts and applications. Cambridge University Press, UK. Pp. 678.

Bolin B. (ed.) (1986). The greenhouse effect, climate change and ecosystems. International Council of Scientific Unions Report, SCOPE 29, Wiley, Chichester.

Chambers, J.Q., Higuchi, N., Shimel, J.P., Ferreira, L.V. and Melack, J.M. (2000). Decomposition and carbon cycling of dead trees in tropical forests of the Central Amazon. Oecologia 122:380-388.

Caspersen, J.P., Pacala, S.W., Jenkins, J.C., Hurtt, G.C., Moorcroft, P.R., and Birdsey, R.A. (2000). Contrributions of land-use history to carbon accumulation in U.S. forests. Science 290, 1148-1151.

Chávez Cortés, J. M. and Trigo Boix, N. (1996). Programa de manejo para el parque nacional Iztaccíhuatl-Popocatépetl. Universidad Autónoma Metropolitana, Unidad Xochimilco, México D.F.

Chomitz, K.M. (2000). Evaluating Carbon Offsets From Forestry and Energy Projects: How Do They Compare? Development Research Group. The World Bank, Washington DC.

Clevers, J.G.P.W. and Verhoef W. (1990). Modelling and synergetic use of optical and microwave remote sensing. Report 2: LAI estimation from canopy reflectance and NDVI: a sensitivity analysis with the SAIL model BCRS-report 90-39. Delft, The Netherlands.

Curran, P. (1980). Multispectial remote sensing of vegetation amount. Progression of the Physical Geography 4: 315-341.

Erb R.B. (1980). The large Area Crop Inventory Experiment (LACIE) methodology for area, yield and production estimation: results and perspectives. In: G. Frauysse (ed.) Remote Sensing Application in Agriculture and Hydrology, A.A. Balkema, Rotterdam, pp. 285-297.

Heil G.W. and Gerrits M. (1992). RANK: A carbon cycle model of terrestrial ecosystems: Evaluation of CO_2 sink enhancement for the Netherlands. Report Resource Analysis, Delft, the Netherlands.

Heil G.W., Gerrits M., Janssen L.H.J.M. and Weenink J.B. (1993). A simulation model to evaluate CO_2 sink enhancement - modelling the carbon storage and the carbon balance of forested ecosystems in the Netherlands. Proceedings IPCC workshop: "Carbon Balance of World's Forested Ecosystems: Towards a Global Assessment". 11 - 15 May 1992, Joensuu, Finland.

Houghton A.R. and Woodwell M.G. (1989). Global climatic change. Scientific American. Vol. 260. no. 4: 36 - 44.

Hughes, R.F., Kauffman, J.B. and Jaramillo, V.J. (1999). Biomass, carbon, and nutrient dynamics of secondary forests in humid tropical region of Mexico. Ecology 80: 1892 – 1907.

INEGI, (1978). Huejotzingo E14B42. Carta topografica 1:50,000, Mexico.

INEGI, (1982). Cuautla E14B51. Carta topografica 1:50,000, Mexico.

INEGI, (1983). Mariano Arista E14B32. Carta topografica 1:50,000, Mexico.

INEGI, (1985a). Amecameca E14B41. Carta topografica 1:50,000, Mexico.

INEGI, (1985b). Atlixco E14B52. Carta topografica 1:50,000, Mexico.

INEGI, (1985c). Chalco E14B31. Carta topografica 1:50,000, Mexico.

IPCC, (1997). Revised 1996 IPCC Guidelines for National Greenhouse Gas Inventories. In: Houghton, J.T., Meira Filho, L.G., Lim, B., Treanton, K., Mamaty, I., Bonduki, Y., Griggs, D.J., and Callander, B.A. (Eds.). Intergovernmental Panel on Climate Change. Meteorological Office, Bracknell, UK.

Kirschbaum, M.U.F. (1999). CenW, a forest growth model with linked carbon, energy, nutrient and water cycles. Ecological Modelling 118: 17 – 59.

Kirschbaum, M.U.F. and Mueller, R. (Eds.) (2001). Net Ecosystem Exchange – Cooperative Research Centre for Greenhouse Accounting. Canberra, Australia. Pp. 136.

Kirschbaum, M.U.F., Schlamadinger, B. Cannell, M.G.R., Hamburg, S.P., Karjalainen, T., Kurz, W.A., Prisley, S., Schulze, E.D. and Singh, T.P. (2001). A generalised approach of accounting for biospheric carbon stock changes under the Kyoto Protocol. Environmental Science & Policy 4: 73 – 85.

Lashof, D. and Tirpak D. (eds.) (1989). Policy options for stabilising global climate. EPA draft report to the US Congress.

Lieth, H. and Whittaker, R.H. (1975). Primary productivity of the biosphere. Springer. Ecological Studies Vol. 14, p. 339.

Mohren, G.M.J. and Klein Goldewijk, C.G.M. (1990). CO2FIX: a dynamic model of the CO2 fixation in forest stands. Report 624, 'De Dorschkamp', Research Institute for Forestry and Urban Ecology, Wageningen, The Netherlands. 96p.

Moore, P.D. and Chapman, S.B. (1986). Methods in Plant Ecology. Blackwell Science Ltd.

Pinty, B. and Verstraete M.M. (1992). GEMI: a non-linear index to monitor global vegetation from satellites. Vegetatio 101: 15-20.

Strahler, A.H. (1981). Stratification of natural vegetation for forest and rangeland inventory Landsat digital imagery and collateral data. Int. J. of Remote Sensing 2: 15-41.

Tegart, W.J. McG., Sheldon G.W. and Griffiths D.C. (eds.) (1990). Climate Change. The IPCC Impacts Assessment.

Ter Steege, H. 1994. Hemiphot, a programme to analyze vegetation indices, light and light quality from hemispherical photographs. Tropenbos Documents no 3. Tropenbos, Wageningen, The Netherlands.

Tucker, C.J. and Miller L.D. (1979). Red and photographic infrared linear combinations for monitoring vegetation. Remote Sensing of Environment 8: 127-150.

UNFCCC, (1997). The Kyoto Protocol to the United Nations Framework Convention on Climate Change, UNEP/WHMO.

Walker, B.H. and Steffen, W.L. (Eds.) (1996). Global Change and Terrestrial Ecosystems. Cambridge University Press, Cambridge.

Watson, R.T., Noble, I.R., Bolin, B., Ravirdranath, N.H., Verado, D.J. and Dokker, D.J. (Eds.) (2000). Land Use, Land-Use Change and Forestry. A special report of the Intergovernmental Panel on Climate Change. Cambridge University Press. Cambridge, UK. p377.

Weismiller, R.A. and Kaminsky S.A. (1978). Application of remote sensing methodology to soil survey research. J. Soil & Water Cons. 33(6), 287 - 289.

Woodwell, G.M. (1984). The carbon dioxide problem. In G.M. Woodwell (ed.), The role of terrestrial vegetation in the global carbon cycle. Measurement by remote sensing. London: John Wiley and Sons, pp. 3-15.2

Chapter 8

Strategic Planning of the Iztaccíhuatl-Popocatépetl National Park

Juan Manuel Chávez Cortés, Marta M. Chávez Cortés, Gilberto S. Binnqüist Cervantes, Iván Roldán Aragón, Euridice Leyequien Abarca, Gerardo Romano Delon

GIS laboratory, Departamento El Hombre y su Ambiente, Universidad Autonoma Metropolitanan Xochimilco, México

Key words: AHP, decision-making, GIS, MCDM, Linear programming

Abstract: Land use decision-making is always associated with how to arrive to the decision, which optimises land use in a region. Forestry management as an activity in competition with other land uses, requires to be located in places where it can be appropriately developed without economical and social disadvantages with respect to other activities like agriculture, livestock or urban use. Finding the best area to be dedicated to forestry and, how many hectares to be used for each forestry alternative gives rise to a compatibility conflict between economical and ecological criteria, which must be solved. Therefore, it is necessary to define the best indicators, which permit to reach a feasible solution in order to maximise both land conservation value and the socio- economical benefits of the forestry activities. In this chapter, the authors describe the strategic planning efforts applied to the recreational area of Iztaccíhuatl-Popocatépetl National Park and one of its outlying municipalities. The research was designed for the application of GIS tools and multi-criteria and multi-objective decision-making approaches.

G.W. Heil et al. (eds.), Ecology and Man in Mexico's Central Volcanoes Area, 173–203.

1.   INTRODUCTION

At the end of the 20th century, in the current environmental planning panorama, when the paradigm of sustainability dominates future visions of economic development linked to conservation of natural capital, it is practically impossible to elude its challenges when the environmental components are the targets. One of the most common ways to face them is focusing the design of territorial management plans on sustainability. Developing strategies that drive the desired courses of action into sustainability is another.

Social actors have their own interests associated to their objectives over different time scales, and they expect that the criteria in which they are interested prevail in any decision-making. Actually, one could say that any essay on landscape planning is a problem of multi-criteria and multi-objective decision-making.

For the purpose of actual strategic planning in the volcanoes' area, one can identify at least three main stakeholders involved in landscape decision-making:

- The land owners or owners of natural resources.
- A scientific-technical group formed by experts who analyse and produce information with varying degrees of quality about the landscape and its resources.
- A financial group that consists of economically powerful people and Government sectors supporting regional development projects and with enough financial resources to sustain them.

These stakeholders play their roles in such a way that the relationships between them give rise to three different types of decision-making described below.

Authoritarian decision-making. Decisions are made always by the financial group and land use allocation decisions stem from their projects or from governmental sectors. Owners of land or natural resources have a sole alternative: to adopt the financial group's decisions or to suffer from lack of funds or political support.

The scientific-technical group usually does not take part in a design plan for the money or in decision-making. Very often, they have only been required to justify the different activities of the plans, based on foreign goals. This mechanism of decision-making belongs to the traditional style of planning, where strategies are designed to be carried out without confronting others. Actually, the opponents do not exist, and neither the surprises!

For a long time, this kind of decision processes has prevailed in the volcanoes' area. The results of it are the actual land use pattern and the fast rate of land conversion to urban use. One can conclude that these decisions have depleted the regional environmental heritage.

Technocratic decision-making. This is another common mechanism of local decision-making. It was implemented in the region for the first time in the 80's. According to this style, land use and natural resources management originates from the relationship between the financial and the scientific-technical group. In planning jargon, these decisions are called "*the council of the elders' decisions*". Once more, the benefits of landowners or natural resource owners are sidelined by "*external*" decisions. Many regional development projects have resulted from this planning approach, most of them with financial support but without much social acceptance.

Democratic decision-making. In this approach the decision-making mechanisms arise from an intensive interaction among the three stakeholders described above. The scientific-technical group is in charge of developing alternatives for landscape use, considering the financial and land owners' benefits. The final decision is made as a result of Communication-Coordination-Arrangement (CCA) processes among financial and land owner groups.

From a scientific-technical point of view, this style is more difficult to apply than the other two, because it involves incorporating the inhabitant's opinions and wishes into the construction of the optimal land use alternatives. This decision-making style has no antecedents in the volcanoes' region, yet the authors consider it to be the most appropriate planning style to come up with feasible regional sustainability.

It must be stressed here that intensive CCA processes between two or more stakeholders, as applied in decision-making, is a field of strategic planning. This modifies the sense of planning, because the new target to search for are flexible and adaptable plans, mostly designed to prevent, rather than construct, the future. This is essentially the same approach that Holling and colleagues have promoted under the name of "adaptive management" (Holling, 1978).

Another important and also strategic view is to consider decision making on landscape allocation as a case of conflict of interests. A basis to this is that conflicts arise in direct proportion to the number of stakeholders or interests involved in the decision-making process. Also, discrepancies occur between current land uses (either inherited or historically established) and optimal land uses suggested by scientific and technical methods.

2.   DEFINING THE PROBLEM AND APPROACHING SOLUTIONS

Sustainable development is not a new concept. According to the U.S.A. Centre of Excellence for Sustainable Development, it is the latest expression of a long-standing ethic involving people's relationships with the environment, and the current generation's responsibilities to future generations. For a community to be truly sustainable, it must adopt a three-pronged approach that encompasses economic, environmental and cultural resources. Communities must consider these needs not only in the short term, but also in the long one. Several questions arise form this concept:

-   How can sustainable development be reached?
-   How can the economic, environmental and cultural issues be linked in a long-term outlook?
-   How can regional strategic planning be translated into sustainable development?

In a decisive paper, Braat and Van Lierop (1986) established a series of environmental and resource policies that might help to answer these questions. While searching for good criteria relating to nature and society's complex relationships, these authors studied and comparatively evaluated over 100 economic-ecological models from 21 countries, and designed a general model of the relationships between socio-economic and natural systems. They described them as socio-economic and ecological subsystems.

From this model they derived five main policies for optimising the relationships between both systems.

1. Ecosystem conservation management. If one of the main ethical principles of sustainable development is the intergenerational equity in at least two generations, then the task is to maintain and increase our grandchildren's' environmental capital. This environmental capital is composed by actual ecosystems, with all its natural resources and all man made infrastructures.

2. Economic optimisation management. Another ethical principle of sustainable development is the racial and sexual equity, together with an equal share of opportunities to attain a better quality of life. Another challenge is, thus, the search for economic development, maintaining as a goal a continuous increase in the quality of life of the region's citizens.

3. Sustainable use of resources. This policy is strongly linked with the principle of intergeneration equity, since its main purpose is the use of resources without depletion. It aims for a human population to use renewable resources below their rate of renewal, and looks for alternatives or replacements for non-renewable resources. Another aim is to maximise the social impact derived from the resource usage, maximising its added value.

4. Sustainable use of environmental services. This policy is also linked to the intergeneration equity principle. The main idea here is that the present generation does not exhaust the ecosystem's air and water cleaning capabilities, neither exceeds the transformation rate of other wastes and residues. On the contrary, these capabilities should be maintained or even increased. The difficulty for the accomplishment of this policy is the lack of development of clean technologies and optimal soil use patterns.

5. Total system management (i.e. sustainable development). The integration of all the policies previously described gives rise to an integrated management system involving nature and society. To arrive to a set of policies for environmental capital conservation,

economic development with progressive increase of life quality, use of natural resources without depletion and permanent environmental services can be the best approximation to the environmental utopia of the end of the 21st century: sustainable development.

Criteria and their indicators should support each of the aforementioned policies aimed to assess and achieve sustainable development. In this manner, it could be possible to compare different alternatives for regional development using the same criteria and indicators. Consequently, the main challenge when trying to link sustainable development and landscape planning is to know how to seize the '*good wishes*' and turn them into realities expressed as alternative landscape mosaics.

When the prospective approach is applied to planning, the construction of a target vision involves the design of desirable, rather than possible futures - a utopia always describes a desirable future-. Sustainable development or any of the policies outlined above could be the desirable vision of a future, a goal or a target to trace paths towards or design strategies to arrive at. Design of strategies also happens to be a part of planning.

The *know how* of transforming objective visions into regional land use models, their strategies and tactics, is the subject of strategic landscape planning. Generally speaking, strategic landscape or regional planning is relevant because it permits us to design strategies and tactics to achieve desired visions of the region or landscape in question, and to prevent unwanted futures. Tactical planning means the design of activities necessary to achieve objectives or goals, and when one designs how such activities could be integrated and structured, one is designing strategies.

A strategic question at the start of a regional sustainable development is: what is the size or complexity of the problems that sustainability poses to regional planning? In the volcanoes' region, problems are large and complex because the conservation of its natural capital is being endangered by the following critical factors.

a) The population and economic growth tendencies in the megalopolis will not change in a near future. According to demographic and economic projected data from the Metropolitan Research Program, the Valley of México, the main sub region of the megalopolis, will have 31.500,000 inhabitants by 2010; and 35.800,000 inhabitants by 2020. The economic development is estimated at the same growth rate (Ramirez, 1997).

b) Consequently, the urban growth rate in a regional context will continue to increase. Furthermore, there are two urban pathways surrounding the volcanoes: Chalco-Cuautla and Puebla-Atlixco. Both of them have the two municipalities with the fastest demographic growth rate, Chalco and Puebla.

c) Unfortunately, economic activities are not integrated with the environment; this results in a sub optimal use of natural resources and environmental services in such a way that contamination and resources shortage in the megalopolis is common.

d) Following as in a chain reaction, the pattern of land use change is towards the depletion of forests. Unfortunate consequences to this are the loss of a variety of natural values such as outdoors recreation, stream control, water collection and recharge of underground deposits, air cleaning, removal of suspended particles in the air and uptake of sulphur dioxide, nitrogen dioxide, carbon monoxide, heavy metals and ozone.

e) Finally, there is no moral awareness in social sectors about environmental risks derived from environmental degradation. This is the result of the predominance of purely economic criteria in land use decision-making. Environmental costs are not included in cost-benefit analysis applied to industrial or urban projects and of institutional mechanisms for environmental quality control are fragile and inadequate.

From the previous discussion it should be clear now that community problems cannot easily be addressed by traditional approaches or traditional social decision-making. It is better to use a more collaborative and holistic systems approach because in their nature such problems are diffuse, multidisciplinary, multi-agency, multi-stakeholder and multi-sectorial.

3.   METHODS

3.1   Systems approach and strategic planning

An appropriate methodology for facing complex problems is a combination of the systems approach and strategic planning. The latter one, being the more suitable mechanism for facing and contending situations with a high grade of uncertainty. Its main quality is to prevent rather than to construct the future. Its tools for facing unexpected situations, have gained adepts for the approach in fields such as business, public policies design, conflict management; and of course in landscape and regional planning. In turn, the systems approach is suitable for three main reasons: first, because there is a need for global or holistic visions of landscape planning problems; also because it permits to integrate different problem solving techniques applied to the analysis of complex systems; and in third place because it is an excellent methodological framework for integrating analytical hierarchical process, operations research and geographical information systems tools (Table 8.1).

Table 8.1. Synthesis of the landscape planning process.

Tools	Analytical Hierarchy Process (AHP)	Operations Research (OP)	GIS Modelling (GIS)
Methods	- Experts judgements - Pairwise comparison - Brain storm	- Linear programming - Multi-objective mathematical programming - Hop Skip & Jump	- Multicriteria/Multi-objective - Pairwise comparison - Weighted linear combination - Heuristical compromise solution
Products	Hierarchy of the problem	Optimal and sub-optimal land use solutions	Spatial allocation for a land use scheme
Type solutions	- Which is best land use according to the goal - Which is the best alternative according to each criteria selected - Which is the criteria with more influence over the problem solution	- Which is the best land use scheme - How many hectares must be dedicated for each land use	- Where must the land use alternatives included in the solution be allocated

Land use planning is the process of evaluating different alternatives with the purpose of deciding on the optimum destiny of the land and formulating territorial legal policies. Nevertheless, a wrong decision with regard to the land use has not only a negative effect on the environmental impact but also a high socio-political cost.

Many decision-making processes applied to territorial planning, move only in one direction, following a temporary sequence known as *forward planning* or *desired phase* (see Figure 8.1). For this type of progression, decisive factors of the present time are analysed and projected to define an expected future.

A somewhat opposite approach known as *backward planning* or *expected future*, starts with the identification and valuation of the intermediate factors and results involved in reaching the expected result and from there develops the strategy to follow (Saaty, 1995).

In both planning approaches, decision-making is a complex process because, as a rule, problems related to land use are of a multi-objective and multi-criteria nature. Besides this, it is also common that decisions are made under restrictions such as the lack of technical data, specialised personnel, time and budget as well as a scarce political disposition of the actors involved.

On the other hand, many proposals for land use allocation are rigid and authoritarian, since they are based on a top to down approach and therefore are not flexible enough when the actors involved in the decision making and its resources change. Thus, it is common to find proposals that are technically ideal, but lack social feasibility because they do not come from a participating context.

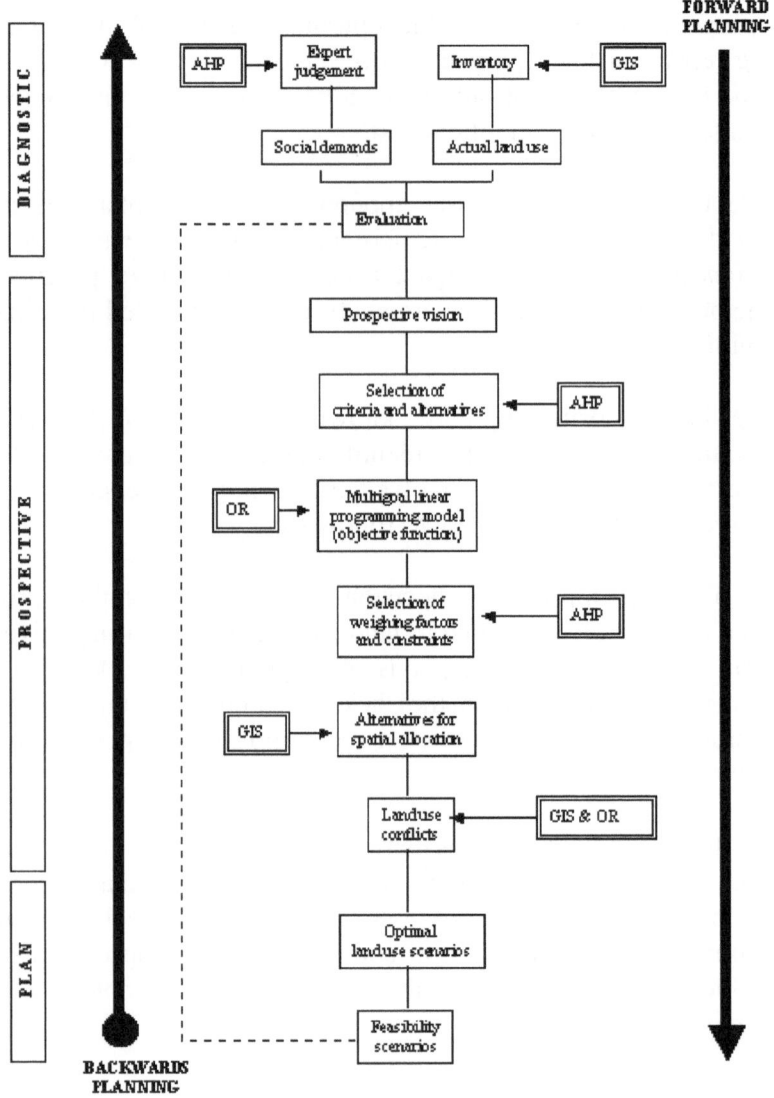

Figure 8.1. An integrating framework for strategic landscape planning

3.2   Analytic Hierarchy Process

To be able to incorporate participation of the actors, an important tool has been developed by Saaty (1980): the Analytical Hierarchical Process (AHP). This tool is focused in developing a structure to the problem and defining criteria organising the problem in a hierarchical manner. It also allows to incorporate experience as a feedback mechanism in the making of decisions.

This technique is a generalisation of the eigenvector method, developed by Saaty during the 70's to determine the weights of a variable through a pairwise comparison matrix. It has been applied mainly in the decision-making of corporate groups and in aspects related to public administration. The scope of these applications include strategic planning, public investment, conflict analysis, marketing, proposal evaluation, equipment purchase and location of resources (Saaty, 1995, Analytic Hierarchy Process Homepage, 1997).

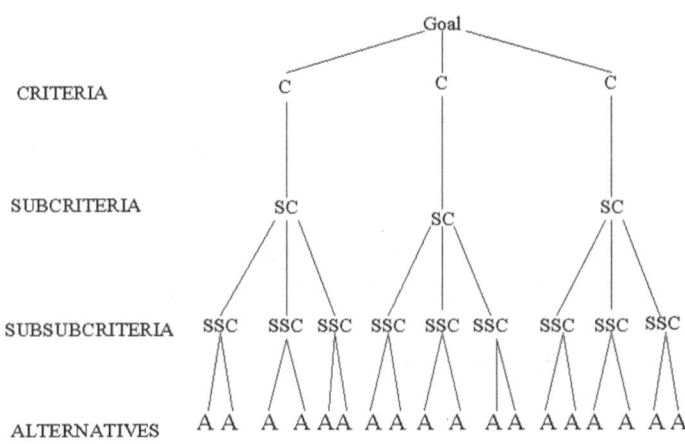

Figure 8.2. Schematic representation of an AHP model.

A distinctive characteristic of the AHP is that it is set up from the principle that to make a decision, experience and knowledge of the people are as valuable as all the data employed. However, all the information should be grouped together through a hierarchical structure or model. An AHP model organises all the different elements of a problem into a structure shaped like a tree (a hierarchy) in which each element of the tree is referred to as *node* (Saaty, 1990). The goal of the problem is known in the higher level; below it, in an intermediate level, the criterions and sub-criterions that are the scenarios, actors and objectives of the problems are found, and in the inferior level the alternatives of solution are found (Figure 8.2).

3.2.1   Scales and Comparison Matrixes

When elaborating a pairwise comparison matrix using the eigenvector method, the relative weights (of importance or preference) for the elements to be compared are established. These weights represent the priorities that should be taken into consideration by a decision-maker when choosing in each node of the hierarchy with respect to the goal.

The matrixes of the AHP are symmetrical, for this reason the relative weights of the elements can be calculated in the following manner:

$$A*| = \lambda max*|$$

Where A is the observed pairwise comparison matrix, λmax. is the maximum eigenvalue of A, and | is its eigenvector, which constitutes an estimate of the relative weights.

Once all the comparisons are done, it is possible to determine the degree of consistency of the judgements done, indicating the probability held by the values in the matrixes. In this sense, Saaty suggests that comparison matrixes where inconsistency values are greater than 0.10 should be re-evaluated.

3.2.2   Synthesis of the Hierarchy

Once all the alternative comparison matrixes have been processed obtaining the weights derived from the different judgements, it is possible to calculate the final priorities that show how the weights of each alternative and criterions with regard to the decision factors and to the goal are ordered.

The results of this process helps on one hand to estimate the final weights of the criteria with respect to the goal of the problem and on the other hand, to choose for the alternative(s) that better suit the considerations of each criterion. The latter is the final result that the decision-maker has for the solution to the problem.

3.2.3   Considerations for the Method in Landscape Planning

The conflicts over the land use can be solved in different ways, may it be by imperative decisions without consulting the affected people or using participatory planning techniques for scenario construction. In this sense, the Hierarchical Analytical Process stands out as a tool to be applied in the land use planning, in as much as it is mainly a method that facilitates the multi-disciplinary analysis, since it is not only a brainstorming of ideas on the structure and solution to the problem, and allows the transformation of these ideas into measurable criteria. Also, given its participating character it gives place to the discussion of criteria and alternatives over the different considerations or disciplinary weights that can happen in the planning of the problem as well as in the generation of the partial or total solution alternatives.

On the other hand, AHP incorporates and considers not only quantitative criterions as the land slopes, vegetation types coverage, the number of employment's generated, the rate of demographic growth, the energy consumption per capita, or the agricultural productivity, but it also considers intangible criteria where an absolute value does not exist. Such is the case of relative criteria such as the preferences or importance for an specific land use, the social acceptation to determinate initiatives of the law or the cultural importance of a place for some ethnical group. Relevant aspects that in a given moment could make a project feasible or not and that with other methods are hardly measurable.

Finally, the hierarchies are very flexible structures that allow to be modified and adapted to unexpected events, such as the change of the actors involved or the variability in the weight of their judgements.

3.3   Operations' research tools usage on landscape planning

3.3.1   Optimisation applied to land use decision-making

Operations' research is a discipline that uses mathematical methods to argument decision-making in any goal-oriented type of human activity. Among its techniques, multi-objective programming has had special relevance on landscape planning. This technique, mainly based on linear programming, makes simultaneous optimisation of several development goals possible, such as maximising profits or ecological value while minimising soil erosion or investment costs. Through linear programming, a solution that best satisfies development goals is found.

In the case of landscape planning, the decision-making problem consists of determining an optimal land use reallocation choosing among different potential land uses in a region with a specific land use. An optimal reallocation is that which performs an objective better than any other possible one, with respect of one or more criteria. The optimal solution defined will depend of the selection, measurement and emphasis of multiple evaluation criteria (Ridgley, 1995).

The outcome from a previous quantitative evaluation of potential land-uses is the input data of the model used to this purpose. The result is a scheme that defines the best combination of land uses and the optimal area to be dedicated for each one, accounting for social, economical and ecological constraints.

3.3.2   Linear Programming

Linear programming (LP) is a mathematical procedure to determine the optimal assignment of limited resources (Schrage, 1991), i.e., it is a procedure to solve optimisation problems. The technique is far from new, being fashionable in the 60's and 70's, with application to problems in areas so diverse as banking, education, and transportation. In recent years, the need for defining land-use allocation and for rural development planning revived interest on this tool in the 80's.

A simple land use problem.
Given 100 ha of land, the alternative is to use them for grassland or forest. For each ha used for grassland, 300 ton of sediment/yr. are delivered, 100 mm of water/yr. are caught and $200 of profit/yr. are made. On the other hand, for each Ha used for forest, 100 ton of sediment is delivered, 50 mm of water are caught and a profit of $150 is made in one year. If a profit of at least $15000 is to be made in a year, limiting sediment delivery due to soil erosion to no more than 15,000 ton/yr. while producing as much runoff as possible, how many Ha should be used for grassland and how many for forest? Clearly this is a problem with a single objective and three restrictions, which can be expressed qualitatively as a LP model as follows:

Maximize runoff
Subject to
(# Ha grassland)+(# Ha forest) <= 100 Ha
($/Ha grassland)(# Ha grassland)+($/Ha forest)(# Ha forest) >= 15,000
(Ton/Ha grassland)(# Ha grassland)+(Ton/Ha forest)(# Ha forest) <= 15,000 Ton
runoff = (mm water/Ha grassland)+(mm water/Ha forest)(# Ha forest)
Quantities to be determined (decision variables): # Ha grassland, # Ha forest.
This statement can be transformed to its quantitative form, giving:

Maximize Z = 100 X1 + 50 X2
Subject to:
    X1 + X2 <= 100
    200 X1 + 150 X2 >= 15000
    300 X1 + 100 X2 <= 15000
    X1 >= 0
    X2 >= 0

X1 and X2 are the unknowns. The grassland/forest land-use problem seen as a multi-objective LP model. A multi-objective linear programming model helps land-use decision-makers to solve situations involving multiple goals and objectives. It makes possible maximization and minimization of more than one development goal, accounting at the same time for constraints specific to the area of study. The land-use assignment problem outlined above would be stated as a multi-objective problem if, for example, we were interested on optimizing not only runoff, but profits and sediment delivery as well, all the while not using more land than is available. Qualitatively stated as a LP model it would appear: Maximize runoff, Minimize erosion, Maximize profits, Subject to: used land <= available land. Now, quantitatively expressed:

Maximize Z1 = 100 X1 + 50 X2
Minimize Z2 = 300 X1 + 100 X2
Maximize Z3 = 200 X1 + 150 X2
Subject to:
X1 + X2 <= 100
X1 >= 0, X2 >= 0

Figure 8.3. Example of Linear Programming problem

When stating a linear programming problem, two classes of objects can be distinguished: limited resources such as soil, production capability of an industrial plant or a store's sales force; and activities such as producing steel or the use of soil (e.g. Figure 8.3). Each activity uses resources or -possibly- contributes some amount of them. The idea is to determine the best combination of activity levels without requiring more resources than those currently available. In this type of problems, the decision-maker seeks to maximise (usually profits or income) or minimise (usually costs) some function of the variables involved. This function is known as the objective function (Winston, 1994).

3.3.3   The Hop Skip & Jump method for generating alternatives

Landscape planning problems are so complex that not all of its relevant aspects can be included in a mathematical optimisation model. This is because such aspects may have a qualitative nature, some may be unknown or not mentioned by involved entities.

For this reason, there are multiple solutions for a mathematical formulation, which are very similar with respect to modelled objectives but drastically different one of each other with respect to the decision variables included. That is why a potentially important role of the optimisation models is to provide a number of alternatives that are feasible and good enough with respect to the modelled objectives, and different in relation with the decisions they specified (Brill *et al.* 1982).

The Hop Skip and jump (HSJ) method is designed with this purpose. This method permits to generate and to identify a variety of alternatives as wide as possible. Increasing the variety of options is possible when inferior solutions in the model are considered. This is very useful to reach a general agreement in conflict situations.

The method consists of finding a non-dominated solution through any optimisation method (e.g. compromise programming technique). After that, to attain an alternative solution solving a linear programming model in which, the objective function is the summation of all decision variables not included in the first solution and the goals become restrictions. After that, consecutive iterations must be done, always including into the objective function all those decision variables not considered in the previous solutions. The process will stop when the differences between solutions are negligible or when there are no more decision variables previously included in the model.

3.4   Geographical information systems applied to landscape planning

Landscape planning has as its fundamental objective the harmonic organization of terrestrial space, based on three complementary criteria relating human societies to their environment: natural resources, users and space. It is thus a complex notion, centred on the concept of geographical space and on the distribution of human activities. In the same way, landscape-planning practice requires studies that include three main components elaborated consecutively: a diagnostic (inventory and assessment); a prospective offering several landscape planning alternatives (generation of scenarios) and a guiding scheme for territorial planning (compromise or consensus scenario) (Zinck, 1996). Regardless of the component being undertaken, decision-making is a process common to all. Whether to determine land suitability for several uses, to generate different land use schemes meeting outlined objectives; or to find a compromise scenario according to its impact, a choice between different alternatives must be made.

To select the best alternative, one or more goals and a number of constraints must be taken into account. They should be measurable and comparable by a choice function, since they will be the evidence that will support the decision.   These last statements make up the core of multi-criteria/multi-objective analysis for problem solving (MCDM). This approach classifies decision-making in all spheres of human activity into four categories according to the relation between the number of objectives and criteria considered.

In the case of landscape planning, examples of this classification are the following:

- One objective-multiple criteria. For example, identification of potential erosion based on the universal soil loss equation.
- Multiple objectives-one criterion. In practice, it is not possible to assess a multi-objective problem using just one criterion. This combination is thus discarded.
- Multiple objectives - multiple criteria. For example, development of a zoning scheme for a coastal lagoon with different purposes based on a variety of criteria.

The inclusion of this analysis approach into GIS technology involves the incorporation of different model types; assuming multiple land use potential maps as the objectives to meet; information layers as the criteria, and decision rules as the objective function.

3.4.1   Spatial information and GIS

The examples in the previous section underscore the spatial nature of landscape planning. This is why spatial information is the backbone to support decisions about land allocation. Nevertheless, procedures for the complete inclusion of this information into a decision making process have been severely limited in the past.

Traditional statistical approaches seek to constrain the spatial variability of data obtained through geographically stratified samplings, reducing thus the spatial complexity of data. However, complex spatial patterns and their interactions are precisely what natural resource managers are most interested in. Furthermore, while this approach preserves the quantitative aspects of data, necessary for most decision-making models, it lacks spatial continuity. On the other hand, the pure map-making approach is spatially precise but limited, due to its non-quantitative nature and its intensive labour. In contrast, GIS technology provides the means for quantitative modelling of entity relationships, accounting for their heterogeneity and spatial continuity. It is in this analytical context that mapped data can truly become spatial information for an effective decision-making (Berry, 1987).

In agreement with Aspinall and Pearson (1993), and Schmoldt *et al.* (1994), GIS applied to landscape planning necessarily includes a modelling approach to territorial usage, towards the evaluation of a plan, policy or directive in terms of both its direct and indirect consequences on the biophysical and socio-economical environment. At the current state of models, plans, policies and directives can be tailored to produce the desired effects, contributing directly to the decision making processes and policies design. There are four main kinds of models available that are suitable for inclusion in GIS technology: rule-based models, also known as spatial coexistence models; knowledge-based models, or deductive models; spatial-inductive models and geographical models.

Rule-based Modelling: Rule-based modelling on a GIS makes extensive use of map overlay and includes what is known as sieved mapping, a term commonly used in Planning. According to Eastman (1993), these procedures are termed Multicriteria Evaluation Techniques (MET) and among them models of a binary nature can be found, in whose applications the GIS user seeks binary answers (yes/no) to her queries, as a function of the spatial coexistence of given attributes. Within this approach, only locations meeting all criteria selected for the evaluation may be chosen, and the risk of not finding any sites that fulfil the requirements is thus accepted.

This type of evaluation imposes heavy restrictions, given its deterministic character. There are other techniques where it is possible to evaluate the alternatives quantitatively and qualitatively, while taking into account site characteristics (uses and/or land suitability). Such measurability of possibilities is of paramount importance in Territorial Planning, where decisions must be made according to multiple, and very frequently conflicting, objectives. Rule-based models can be modified along these lines, to allow the use of weights applied to the different information layers to be overlaid and to the several elements represented within each layer

Deductive Modelling: Deductive models assume there is enough knowledge on the processes to be modelled to establish governing equations, and that information about the variables of these equations is available in the GIS data base. From this basis, equations are written and relations developed outside the GIS and linked to a certain territory through the geographical database. The main contribution of these models is for impact assessment.

Inductive Modelling: Inductive spatial modelling uses a process in which, from specific facts, general relationships are found among the geographical database items. The relationships help to make a decision using rule-based modelling. Examples of these models are the indices used to evaluate ecological variables.

Geographical Modelling: Finally, geographical modelling describes spatial patterns in the database as a function of their location. A vegetation map, for instance, is a geographical model describing spatial patterns for different types of vegetation cover on a section of land.

3.4.2   Planning the Landscape

All models mentioned so far can be incorporated to the different stages of territorial planning through one or several stages in the process.

Diagnostic Stage. Geographical modelling can represent spatially an inventory of available natural resources (such as soil or hydrology) or the spatial location of land suitability for multiple uses. In a similar way, inductive modelling is a vehicle to generate information layers that will be the criteria to evaluate land suitability (a slope map, a distance to water map, etc.) On the basis of these information layers, rule-based modelling can control their combination, acting as a heuristic decision function. The result will be a land suitability map.

Projection-prospecting stage. At this stage, deductive modelling will create maps showing areas that will most probably be subjected to land use change, or maps depicting the impact of a specific land use change. In its turn, inductive modelling together with geographical modelling will serve to obtain information layers that will feed the deductive model. This information, summed up in tables, statistics and graphs for a number of environmental units, provides direct feedback to a policy. With outputs such as these, the consequences of land uses may be discussed, increasing significantly the probability of obtaining the desired solution by originating different scenarios. At this stage, rule-based modelling has special importance when there are spatial conflicts in the production of scenarios. The weighted linear sum, for example, ranks usage preferences in regions with multiple vocation.

Guiding Scheme. In the context of strategically regional planning, focused on territorial management, only one scenario, which is the result of consensus or compromises supported on complementary scenarios, is applicable. Consequently, the different generated alternatives must be evaluated in relation to their environmental, social, economical, political and organizational feasibility. The creation of possible scenarios showing the impacts of policies proposed by the different actors involved in planning, gives decision makers a way to identify conflicting areas, negotiate between alternatives and generate a consensus solution over territorial usage (Zinck, 1996).

One of the great advantages of modelling is its theoretical ability to forecast the events resulting from the action of a system. For the researcher, a model is the means to experiment and verify his hypotheses by selective modification of the model and the input data. To the land manager, it is the means to simulate the transformations a territory will undergo when its initial condition is disturbed (prevention model).

Another interesting point of simulations is the possibility of confronting their results with the real world, so that the experience gained can be reintroduced into the model. Reintroduction should be performed in a very short time to keep transformations from becoming irreversible.

4. CASE STUDY. MULTICRITERIA ANALYSIS AND GEOGRAPHICAL INFORMATION SYSTEMS FOR DECISION-MAKING IN FORESTRY MANAGEMENT

4.1 Introduction

The following section is an example of the application of the framework depicted in the previous sections of this chapter. This case study is to perform a land-use planning process in the Ecatzingo de Hidalgo municipality on the Southwest of the Popocatépetl using forestry management criteria as the backbone for management proposals. The problem was undertaken with a combination of three multi-criteria/ multi-objective analyses techniques (see theoretical background).

First of all, a problem model was structured through the analytical hierarchy process in order to define the decision factors, criteria and alternatives that describe it. Then, using the pairwise comparison method (Saaty, 1990), the weighing values of the alternatives respect to the criteria were calculated. Afterwards, a multi-objective linear programming model was constructed considering the alternatives and including the weights mentioned before. Finally, following a multi-objective/multi-criteria approach using GIS technology, land-use suitability maps were elaborated and the optimal spatial allocation of the programming model results were found. As a result, based on forestry criteria, one optimal and two sub-optimal land use scenarios were obtained.

4.2   Methods and Results

4.2.1   Analytic hierarchy process

Twenty-two land-use change alternatives were determined from a land use and vegetation map of the Iztaccíhuatl-Popocatépetl National Park elaborated by Van de Poll (1995). (Table 8.2 and 8.3).

Table 8.2. Land use classes

Proposed land uses			
Actual Uses		Potential Uses	
Pine forest	Pinus hartwegii forest with tall grass	Conservation forest	Pinus spp. or Abies religiosa forest in shifting mosaic steady state, it means a forest, which can be present in a wide scale all the development states
Open Pine Forest	Pinus leiophylla, Pinus montezumae, Quercus spp., and Abies religiosa forest with low tree density, shrubs and grass	Restoration areas	Eroded areas and old agriculture lands for erosion control and reforestation
Fir Forest	Abies religiosa forest up to 2800 masl, mainly in brooks	Agroforestry areas	Trees in live fences around agriculture areas
Mixed Forest	Pinus spp. (pine) and Quercus spp. (oak) and Arbutus mexicana shrubs and grass	Forest grazing areas	Grassland areas with open pine forest and fire control areas
		Managed forest	Pinus spp. forest and Abies religiosa forest with a selection system forest management

Table 8.3. Land use change classes

Land use changes	key
Mixed forest to managed Pine forest	MPM
Mixed forest to conservation	MC
Mixed forest to grazed forest	MSP
Agriculture to Agroforestry	AAS
Agriculture to Restoration	AR
Agriculture to Conservation	AC
Open Pine forest to managed Pine forest	PAPM
Open Pine forest to restoration	PAR
Open Pine forest to conservation	PAC
Open Pine forest to grazed forest	PASP
Fir forest to conservation	OC
Fir forest to grazed forest	OSP
Fir forest to managed Fir forest	OOM
Grassland to agriculture	PSA
Grassland to restoration	PSR
Grassland to conservation	PSC
Grassland to grazed forest	PSSP
Grazing land to Agroforestry	PSAS
Pine forest to grazed forest	PSP
Pine forest to conservation	PC
Pine forest to managed Pine forest	PPM

A hierarchy model was constructed by decomposing the original decision problem into interrelated decision elements (Figure 8.4). Starting with the goal at the top of the hierarchy, it proceeds through the intermediate levels with the decision factors and it continues with the criteria that influence the decision. How specific these criteria are is a matter directly related with the depth of the levels. At the end of the hierarchy, decision alternatives - the land use changes- are stated. In order to compare the solution alternatives, intensity scales were constructed, one for each criteria. These scales were built using actual standardised data or, when data was not available, through relative measures - i.e. new scales development-. When it was the case, based on the experts' knowledge, extreme nominal values were established and from these, medium values were estimated; these values served to build the scale.

Figure 8.4. Hierarchy model

The scales were designed from high to low values and the numerical values assigned to the nominal ones was carried out applying a rating scale approach (Peterson *et al.* 1994). When the scales were built from actual data, the value of each criteria with respect to each alternative was established (e.g., the stand volume generated for a specific land use or the execution costs for a land use change). These values were finally standardised and from them the intensity scales were constructed; alternatives were then rated using those scales. Finally all the alternatives were compared against the same standards and rated by them. In this way, AHP allowed us to obtain numeric weights or values that reflect the relative degrees in which each alternative reached the overall objectives.

4.2.2   Multi-objective Mathematical Programming

Goals

Two goals were stated to be optimised: the ecological value and the economical value. With the land use model, the optimal combination of hectares that must be dedicated to each of the proposed land use changes and that maximises both environmental capital and economical benefits produced by the activities was found.

Constraints

The following constraints were considered in the land use model:

Total area available to each present land use capable of changing to a potential land use should not be exceeded.

land use j <= # hectares available of land use *i* capable of changing to land use *j*

Application of the model

The land use model was based on activities (land uses) that contributes to the formulated goals and, at the same time, deplete available resources (land). The formulation of the model included the construction of a payoff matrix consisting of 22 land use changes (rows) and 8 criteria (columns). The payoff matrix contained the coefficients of the model, produced by the AHP evaluation. Part of such matrix is shown in Table 8.4.

The optimisation model was solved using the Hop Skip and Jump (HSJ) method. The first step was to find a non-dominated solution through compromise programming technique. The second step was to attain an alternative solution solving a linear programming model where the objective function is the summation of all decision variables not included in the first solution and the goals become restrictions. After that, consecutive iterations were done, always including into the objective function all those decision variables not considered in the previous solutions. The process was stopped when the differences between solutions were negligible.

Table 8.4. **Payoff matrix.** ++ Non-value Land Use Change

Land use changes	Species density	continuity	Vertical stratification	Canopy cover	Standing volume	employment	income	Costs
MPM	0.712	0.677	0.655	0.089	0.017	0.023	0.008	0.006
AAS	0.112	0.126	0.168	0.021	0.092	0.128	0.069	0.158
PAPM	0.712	0.677	0.655	0.097	0.035	0.023	0.008	0.006
OOC	1.000	1.000	1.000	0.088	0.055	++	++	++
PSAS	0.112	0.126	0.168	0.021	0.092	0.128	0.069	0.158
PSP	0.27	0.191	0.168	-0.003	-0.184	0.078	0.177	0.061
PC	1.000	1.000	1.000	-0.008	++	++	++	++

Some of the solutions obtained by the optimisation model are shown in Table 8.5. Ten solutions were originally calculated but solutions 5,7 and 10 were found to be identical to 3,6 and 8 respectively. This is why they are not included here.

Table 8.5. Optimization model solutions. Note: all solutions have a margin of 30 and 35% with respect to economical and ecological restrictions

Key	Solution1	Solution2	Solution3	Solution4	Solution6	Solution8	Solution9
MPM	503 ha		474 ha		302 ha		
MC			68 ha				
MSP	39 ha	542 ha		542 ha	240 ha	542 ha	542 ha
AAS	401 ha			32 ha			
AR							
AC		442 ha		410 ha	442 ha	442 ha	
AA	41 ha		442 ha				442 ha
PAPM	437 ha		437 ha		437 ha	437 ha	278
PAR							
PAC							159 ha
PASP		437 ha		437 ha			
OOM	112 ha		514 ha		1048 ha	849 ha	
OOC	1046 ha	1158 ha	644 ha	1158 ha	110 ha	309 ha	1158
PSA							71 ha
PSR							
PSC							
PSPS	71 ha	30 ha	71 ha	71 ha	71 ha	71 ha	
PSSP		41 ha					
PSAS							
PSP	627 ha	581 ha	346 ha	540 ha	627 ha	464 ha	255 ha
PC		46 ha		87 ha			372 ha
PPM			281 ha			163 ha	

4.2.3   Geographical Information Systems

To find the optimal allocation of the amount of hectares assigned to each land use change, a multi-criteria/multi-objective approach was used in the application of geographical information systems. The process consisted of generating a suitability map for each land use alternative included in the optimisation model solution, allocating the amount of hectares of each alternative in the best locations within the suitable area and solving the spatial conflict between alternatives.

To construct suitability maps, meaningful biophysical criteria (factors and restrictions) were established. The factors considered were terrain slope and road distances. The only restriction was that the available land for a current use should be compatible with the proposed change. A map for each factor and the constraint were built.

The Eigenvector method was used to calculate weights that reflect the relative influence of the factors over the land suitability for each alternative (Saaty, 1990, cited by Eastman, 1995). Finally, the factors were combined through a weighting sum process and the resulting map was overlaid with the restriction map using a multiplying operation (Figure 8.5).

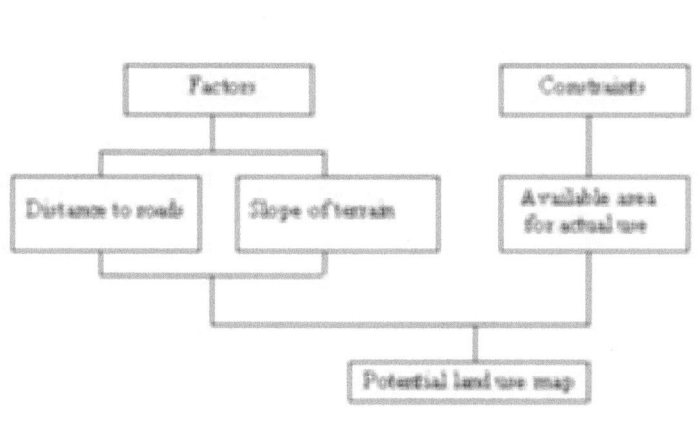

Figure 8.5. Cartographic model for potential land use map

Following the creation of suitability maps, alternative land uses with a spatial conflict were identified. To solve these conflicts, the alternatives were ranked based on expert preferences. The final products, were three land use scenarios: one that represents the optimal solution 1 and two other corresponding to the sub optimal solutions 3 and 6 (Figure 8.6 and 8.7).

OPTIMAL LAND USE SCENARIO OF ECATZINGO DE HIDALGO, EDO. MEX.

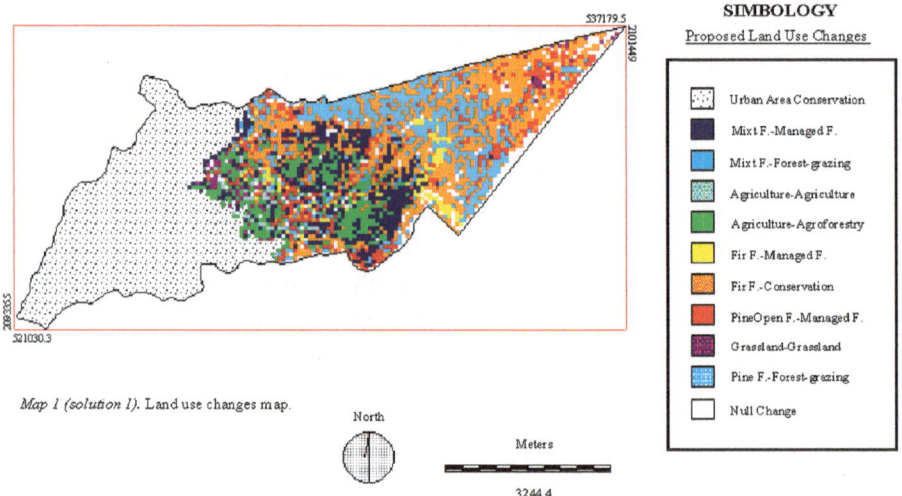

Figure 8.6. optimal land use scenario of Ecatzingo de Hidalgo, Edo. Mex.

SUBOPTIMAL LAND USE SCENARIO OF ECATZINGO DE HIDALGO, EDO. MEX.

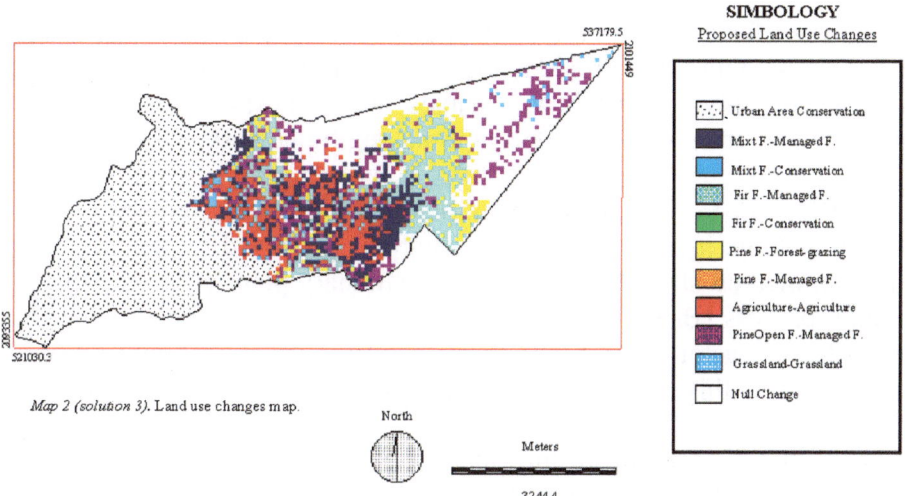

Map 2 (solution 3). Land use changes map.

SUBOPTIMAL LAND USE SCENARIO OF ECATZINGO DE HIDALGO, EDO. MEX.

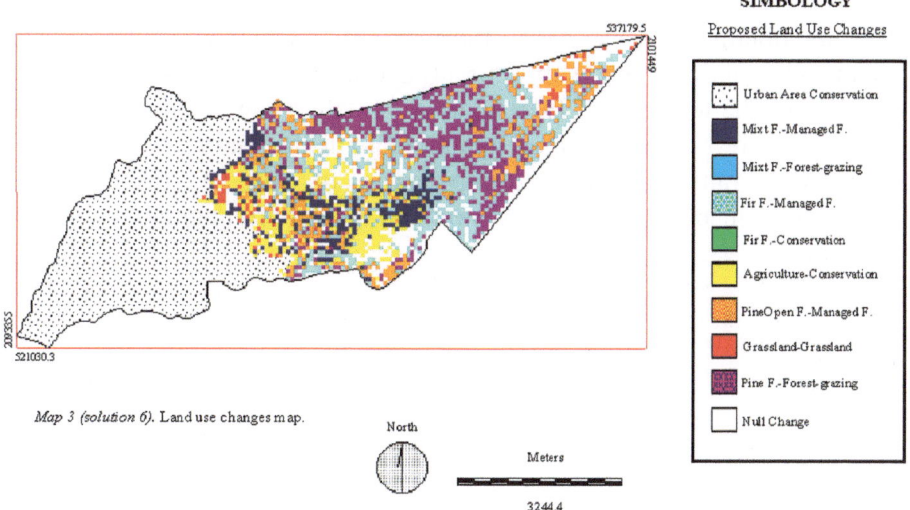

Map 3 (solution 6). Land use changes map.

Figure 8.7. Sub-optimal solutions

5.   CONCLUSIONS

Analytic Hierarchy Process (AHP), can be regarded as a methodology for decision-making that provides tools to construct a decision framework for problem solving that also enables the inclusion of judgements (qualitative criteria) within the decision-making framework.

With respect to the use of operations research as a scientific tool for decision-making, it must be said that solutions generated by the Hop Skip and Jump method should be seen as a preliminary plan within the overall planning process (Brill, 1982) presumably to serve as a starting point for the different actors to base their actions related about resource management; in this case, specifically land use and forestry management.

About the Geographical Information Systems approach, the spatial analysis used to determine land use suitability yielded information to select uses that reduce environmental conflicts and increase social and economical benefits. Different courses of action could be represented spatially. Consequently, this kind of analysis should be included in decision-making as it increases human abilities to generate and to assess alternative solutions for real problems.

Results obtained in the case study met the demands posed by the problem. The optimal and sub-optimal alternatives represent a feasible basis to order the territory of Ecatzingo de Hidalgo, suggesting productive activities that are both economically and ecologically viable. Furthermore, the scenarios rank highly the importance of forestry resources as self-generating media, producing economical and environmental benefits at the same time.

REFERENCES

Analytic Hierarchy Process Homepage. 1997. Http://ahp.net/methodology/ahp/ahp.htm

Aspinall, R. J. and Pearson, D. M. (1993). Data quality and spatial analysis: analytical use of GIS for ecological modelling. Proceedings of the Second International Workshop/Conference on Integrating GIS and Environmental Modelling, Breckenridge, September 1993. NCGIA.

Braat, L. and.Van Lierop, W.F.J. (1986). Economic-Ecological Modeling: An Introduction to Methods and Applications. Ecological Modelling. 31: 33-44.

Berry, J., 1987. Fundamental operations in computer-assisted map analysis. International Journal of Geographic Information Systems, 1:2, 119-136.

Brill, E. D., Jr., Chang, S.Y. and Hopkins, L.D. (1982). Modeling to Generate Alternatives: The HSJ Approach and an Illustration Using a Problem in Land-Use Planning. Management Science 28(3):221-235.

Eastman, J.R.. (1993). Decision Theory and Gis. Proceedings Africa GIS'93 UNITAR/UNSO/SSO, 45-64.

Holling, C.S..(1978). Adaptive Environmental Assessment and Management. John Wiley. New York. 377 pp.

Norton, G.A. and Walker, B.H. (1982). Applied Ecology: Towards a Positive Approach. I. The Context of Applied Ecology. Journal of Environmental Management. 14: pp. 309-324

Ramírez, B. (1997). Diagnóstico Integrado. In: Eibenschutz, R. Coor. , Bases para la Planeación del Desarrollo en México. Toma II: Estructura de la Ciudad y su Región.1997. Universidad Autónoma Metropolitana-Xochimilco, Grupo editorial Miguel Angel Porrua. México. 477 pp.

Peterson, D. L., Schmoldt, D.L., and Silsbee, D.G. (1994). A case study of resources management planning based on multiple objectives and projects. Environmental Management 18:729-742.

Ridgley, M. (1995). Determining rural land use goals: Methodological primer and application to agro forestry in Italy. The Environmental Professional. 17: 1-15.

Ridgley, M and Chai, D. (1990). Evaluating potential biotic benefits from Conservation: Anchia 2line ponds in Hawaii. The Environmental Professional. 12: 214-228.

Saaty, T.L. (1980). The Analytic Hierarchy Process, McGrawHill, New York. 287 pp.

Saaty, T.L. (1990). Multicriteria Decision-Making: The Analytic Hierarchy Process. RWS Publications, Pittsburgh. 268 pp.

Saaty, T.L. (1995). Decision-Making for Leaders. RWS Publications, Pittsburgh. 315 pp.

Schmoldt, D. L., Peterson, D.L. and Silsbee,.D.G. (1994). Developing inventory and monitoring programs based on multiple objectives.Environmental Management 18:707-727.

Schrage, L. (1991). LINDO User's Manual and Software, The Scientific Press, South San Francisco.

Van de Poll, H.M. (1995). Remote Sensing Based Vegetation mapping of the National Park Iztaccíhuatl - Popocatépetl, México. Faculty of Biology, Universiteit Utrecht, report no. 951102.

Winston, W.L.1994. Investigación de Operaciones. Aplicaciones y Algoritmos. Grupo Editorial Iberoamericana, México.

Zinck, A. 1996. La información edáfica en la planificación del uso de las tierras y el ordenamiento territorial. XIII Congreso Latinoamericano de las Ciencias del Suelo.

Chapter 9

Ecology and Man: Synthesis

Gerrit W. Heil*, Roland Bobbink**, and Nuri Trigo Boix***

*Department of Plant Ecology, ** Department of Landscape Ecology, Faculty of Biology, Utrecht University, The Netherlands; ***Departamento El Hombre y su Ambiente, UAM-X, México

Key words: human impact, vegetation, wild life, modelling system dynamics, sustainable development, natural resource management

Abstract: The general public has recognized that environmental problems are likely to cause the current decline of nature. Because of land use intensivation, but also because of fragmentation, less area has become available to plant species. However, the effect of management on natural ecosystems is not always clear. Reflecting this concern, in 1991 the development of a management plan for Iztaccíhuatl and Popocatépetl National Park and its surroundings has been started. A vegetation survey of the Iztaccíhuatl and Popocatépetl volcanoes shows the underlying processes that determine the distribution of the various vegetation communities. Although it has been proved that wildlife, especially mammals and birds, may play an important role in ecological, economic and social values, as yet little research has been done in the area to include wildlife in land use management plans. The results of this book show that Remotely Sensed information could be successfully used as a primary data source for environmental planning. Following a set of guidelines in order to minimize adverse effects on nature is relatively simple, but to manage land use for purposes of nature restoration is much more complex. Here we come to an approach of Natural Resource Management, which should make clear what steps have to be taken.

G.W. Heil et al. (eds.), Ecology and Man in Mexico's Central Volcanoes Area, 205–219.

1.   INTRODUCTION

The richness of vegetation types in México is overwhelming. Most of the major biomes are found in México. However, changes in land cover and the pattern of these changes are related to the observed changes in land use and the increase of disturbances in the Iztaccíhuatl and Popocatépetl volcanoes area. The ecological importance of these disturbances for the structure and function of the natural (forest) ecosystems are shown in the different Chapters. Management strategies are scarce so that the process of reduction in biodiversity continues. The question at the start of the research underlying the different Chapters of this book was: what is the size and complexity of the problems that management poses to regional planning (Chávez and Trigo, 1996)? The problems are large and complex because the conservation of the natural resources is being endangered by different factors. The aim of this synthesis is to draw conclusions about the state of the art in our current knowledge of the interaction between ecology and man in mountainous volcanoes in Central México.

2.   VEGETATION

To be able to design and implement adequate restoration and conservation plans for the Iztaccíhuatl and Popocatépetl volcanoes area, it is necessary to obtain more knowledge of the different vegetation communities occurring in the area as well as of their spatial distribution. For this purpose, in **Chapter 2** the results of a field study has been shown to accomplish a vegetation survey of the Iztaccíhuatl and Popocatépetl volcanoes, which shows the underlying processes that determine the distribution of the various vegetation communities.

The vegetation roughly separates in nine categories:
1. Fir forests: This is a perennial forest of the conspicuous species *Abies religiosa*. The altitudinal range for this community is from 2,700 to 3,550 m. The tree canopy cover is usually dense with relatively tall trees.
2. *Pinus hartwegii* forests: Being the most distinctive of the pine forests, *Pinus hartwegii* is located between the 2,900 and 4,000 m and marks the upper limit of the forest vegetation in the mountains.

3. *Cypress* forests: Although *Cupressus* spp. is strongly associated with *Abies religosa*, in some areas it can be found in pure patches.
4. Oak and alder forests: These broad-leaved forests are located between the 2,350 and 2,800 m. They are found both in deep as in shallow soils.
5. Mountain cloud forest: The mountain cloud forest is a very rare vegetation type, which is present mainly on the west slope of the Iztaccíhuatl volcano in a much-reduced area.
6. Mixed forests: When they are mixed, these forests are named according to the dominant species and this varies according to the altitude as well as changes in the local conditions of humidity, exposition and soil type.
7. Alpine Grasslands: These communities are given many different names: high paramo, alpine prairie or *zacatonal*, a word derived from the aztec word ***zacate***, which refers to those high tall and clumped grasses such as *Muhlenbergia* and *Festuca*.
8. Induced Grasslands: Although there is usually a strong discussion as to when grassland is induced and when it is natural, the characteristic species of the alpine grasslands are usually much reduced in their density in disturbed areas.
9. Secondary vegetation: An increasing vegetation type is the secondary vegetation that is taking over many pine and pine-oak forests because of human activities. When this kind of vegetation is derived from fir, oak or pine forests it is common to find several *Senecio* species dominating.

However, the two main forest communities found in the Iztac-Popo area are *Abies* and *Pinus* forest. From a sub-set of field data, a cluster analysis shows that, depending on the amount of disturbance; the *Abies* associated forest communities can be subdivided into two types (Table 9.1):

AI) a community of herbs, shrubs and small trees: there is local dominance of *Sambucus mexicana* with abundant herb and shrub layer species. Characteristic species for this forest type are *Solanum appendiullatum, Sambucus mexicana* and *Senecio sanguisorbae. Abies religiosa* is absent in this forest type.

AII) "Typical" *Abies religiosa* forest. Forest with tree- shrub-, herb and ground layer species. Characteristic species of this forest are *Alchemilla pinnata, Abies religiosa, Senecio angulifolius, S. toluccanus* and *Symphoricarpus mi-*

crophyllus. Other tree species do occur in this forest type, such as *Cupressus lindleyi* and *Pinus montezumae.*

Table 9.1. Synoptical table of a sub-set of the data of Chapter 2 on (disturbed) forest communities

Plant species	Abies forest			Pine forest	
	AI	AII	BIa	BIb	BII
Solanum appendiullatum	3				
Sambucus mexicana	3				
Senecio sanguisorbae	4	2			
Senecio angulifolius	1	4			
Galeum achenbornii	2	2			
Abies religiosa	-	5			
Alchemilla pinnata		4			
Cupressus lindleyi		1			
Senecio toluccanus		5			
Symphoricarpus microphyllus		2			
Pinus leiophylla			5		
Alnus jorullensis			1		
Stevia spp.			2		
Quercus laurina			1		
Lupinus bilineatus			1	1	
Verbesina virgata			1	1	
Penstemon barbatus				3	
Salix oxylepis				1	
Stevia nepetaefolia				1	
Bouvardia ternifolia				1	1
Oxalis spp.				2	2
Alchemilla procumbens				1	5
Pinus hartwegii					5
Muhlenbergia macrcoura					2
Achillea millefolium					1

Similarly to the *Abies* forest, the main *Pinus* associated forest communities can be subdivided into three types:

BIa) *Pinus leiophylla* forest is characterized by the presence of *Pinus leiophylla.* The vegetation structure is a tree layer varying in density; shrub layer, well-developed herb layer and a ground layer. Other characteristic species are *Lupinus bilineatus, Stevia spp, Verbesina virgata.* Other tree species are present, such as *Quercus spp.* and *Alnus jorullensis.*

BIb) Pinus forest undergrowth community. Very variable tree or small tree layer, shrub layer, well developed herb- and ground layer. Characteristic species are *Penstemon barbatus, Salix oxylepis.* A lot of species are being held in common with community BIa, and also with BII, constituting an undergrowth community in *Pinus* forest in general. *Pinus* species itself are absent in this community.

BII) *Pinus hartwegii* forest community. This vegetation type is open forest, sometimes a shrub layer present with grassy undergrowth of grasses like *Muhlenbergia macroura* and a ground layer. Characteristic species such as *Pinus hartwegii, Achillea millefolium*, and *A. procumbens*. Quite some species occurring in this community are also present in community BIb, such as *Oxalis* spp.

More data on the vegetation composition have to be collected to create a typical vegetation classification (*c.f.* Braun-Blanquet 1964) and to get more insight on the population structure of the different tree species.

3.   HUMAN IMPACT

As in many developing countries, the urban population and its surrounding areas strongly increased in the last half of the 20th century. This certainly holds for the Valley of Mexico; the population of the metropolitan area of Mexico City was just above 1 billion in 1940, and probably around 25 million in the late 1990's. Moreover, the total surface of the urban settlements increased enormously in this period, while the area occupied with agriculture did not decreased. Moreover, the human settlements also spread over the whole region of the Valley of Mexico, researching the foothills of the Izta-Popo area increasingly. In addition, the population pressure increased in recent decades in the Puebla parts of the area, too.

In **Chapter 3** an analysis is given of the changes in natural communities in the Iztac-Popo area in the period 1986 to 1997 using satellite images and ground data of the vegetation. Especially the changes in the density of the different forested areas are treated in detail. The ecological consequences of these disturbances for the structure and function of the natural (forest) systems are obvious. This includes the possible impacts on erosion, and the risks for hazards in the future. The most obvious changes between 1986 and 1997 were observed in the forested parts of the Iztac-Popo area. The following trends were most obvious:

- markedly more very sparse *Pinus hartwegii* stands were found at the highest forest sites in 1997 than in 1986;
- at the lower slopes near the agricultural land use units, the area with dense *Pinus*, *Abies* or mixed forest decreased;
- a large increase in area in shrubs with very sparse tree cover was found in the central-eastern slopes over a large altitudinal gradient.

The deforestation rate (1.2 % per annum) is in line with the overall annual deforestation rates in Mexico (0.55 – 1.30). Besides the reduction in area occupied by forests, fragmentation is also an ecologically important phenomenon.

From the results it became obvious that the decrease in forest density is considerably higher besides trails and small roads, both unpaved. Moreover, dense *Abies* or *Pinus* forest stands are found only on the steeper slopes of the Iztac-Popo. This clearly demonstrates that the observed deforestation is strongly influenced by the accessibility of a forest stand; especially distance to trails for the transport of wood and the steepness of the stand are the most important in this rural mountainous landscape in this respect. Three main activities (or their combination) can be identified as the main causes of this decline in forest cover, viz.:

- strongly enhanced frequency of burning, associated with the strong increase in livestock grazing (cattle, sheep);
- increased logging of trees, especially because of the use of modern cutting and transport ways;
- increased settlements, especially at the eastern part of the volcanoes area.

In Chapter 3 we conclude that the frequent burning, increased grazing pressure and logging activities will obviously lead to bare soil over large stretches in the long term. This will have severe consequences for the erosibility of the slopes, and, finally, on the important water retention capacity of the Iztac-Popo area, as in any mountainous regions.

4.   WILD LIFE

The Ministry of the Environment, Natural Resources and Fisheries, recognises that the offer and demand of wildlife resources and products confers ecotourism and wildlife hiking a potentially important economic role to nature. At the same time, the Ministry emphasises that for these activities money should be generated corresponding to 5% of the total income of the tourist sector in México. Boo (1993) stated that in order to implement an ecotouristic strategy for protected areas, the status of the natural resources has to be established, together with a cost-benefit and an offer-demand analysis of ecotourism. She also mentions the need to have inventories of wildlife places, ecosystems or species that could be tourist attractions.

In **Chapter 4**, we point out the most meaningful bird species for ecotourism as well as the places that can potentially offer great pleasure to the bird watching visitor of the area. This is done through two criteria: the attributes of the birds and those of the most accessible environments. The objective is to delimit the most attractive bird species from the point of view of their visual attractiveness, endemism, abundance, risk category and qualitative appreciation of each species based on the feasibility of watching it in its natural habitat. The species suggested as indicators for ecotouristic interest for each vegetation type are listed. Ten routes along the northwestern part of the Iztaccíhuatl are recommended for bird watching. The region is selected because of its proximity to lodging areas such as Amecameca and Tlalmanalco and also because access to the Popocatépetl area is still risky due to the volcano's low but constant activity.

On basis of the results of Chapter 4, it is obvious that an update of bird species currently active in each vegetation type is urgently needed. On basis of this update a bird's catalogue for environmental education and ecotourism could be created.

In **Chapter 5**, a database is compiled including all mammalian species recorded in the region from 1839 up to 1997. We needed this compilation to understand historical distribution patterns of some taxa, as well as to include data from collections and bibliographic references (see Chapter 5). In addition, field data taken from 1985 until 1997 are also incorporated in the database. All information was then pooled into a final database which included all geo-referenced locations of records as accurately as possible so that links between species records and habitat types was likely to be conducted. The data were then used to build a database of land use types and mammalian records.

To quantify landscape and habitat changes, the two maps of 1986 and 1997 were used for cross tabulation. Major changes from high to medium and medium to low habitat richness classes are happening throughout the whole area. This reduces the resilience of the natural landscapes and increases the threats for most mammalian species since connectivity is important to fulfill daily and yearly mobility. Comparing the fragmentation process between 1986 to 1997 images, it can be observed that only about 20% of the whole area is presently forming continuous habitats, whereas about 80% is already undergoing some form of habitat transformation.

Although it has been proved that wildlife, especially mammals and birds, may play an important role in ecological, economic and social values, as yet little research has been done in the area to include wildlife in land use management plans. Any land cover change impacts in one way or another current mammalian assemblages. The present analysis shows, that especially fir forests, mixed deciduous-pine forests, and alpine grassland communities should have priorities for mammal conservation. If we use mammals as umbrella taxa, natural landscape planning efforts can be undertaken. In addition, elaboration of environmental education brochures about the mammals present in the area will improve the knowledge and sensibility of inhabitants and visitors about the value of the mammal species.

5.   MODELLING ECOSYSTEM DYNAMICS

Dynamics of vegetation occur over a wide range of spatial and temporal scales. The result is that vegetation is patchy on virtually every level of space and time. In vegetation, spatially localised disturbances interrupt assembly processes that would otherwise drive vegetation uniformly toward relatively homogeneous end states. There are three major theories of community assembly, i.e. the deterministic, the stochastic and alternative stable states (ASS). In the deterministic model, a community is seen as the inevitable consequence of physical and biotic factors. In the stochastic model the community composition and structure is essentially a random process. The ASS theory is intermediate between the first two. The importance of disturbance in community dynamics has relatively recently been recognised as a major factor to community assembly. In **Chapter 6** we discuss the potential of a stochastic non-mechanistic modelling approach, and to apply this model to the real-life plant community dynamics of the Izta-Popo National Park in Mexico. The results show that interconnections and feedbacks between two main processes; i.e. community assembly and stochastic disturbance can become visible.

Many countries, lack sufficient information on the actual ground cover of different types of (agro) ecosystems, i.e. on carbon pools in the form of terrestrial ecosystems. Without such land cover data, it is almost impossible to analyse properly the natural C-cycles of these countries. However, such data can be obtained with remotely sensed data from satellite images. In **Chapter 7**, we studied the effects of changes in land cover on the C-cycle and the time-scales over which they occur.

We applied a dynamic simulation model on carbon cycling of terrestrial ecosystems, which has been converted into a spatial (GIS) environment. The results of our case study show a significant correlation between field data on the amount of biomass/carbon of different natural forests and NDVI values from a TM satellite image of the study area. The model has been applied to study the effects of fragmentation on the carbon cycle of the Iztaccíhuatl - Popocatépetl National Park. There is a negative effect on the carbon content in the different soil and vegetation compartments due to the fragmentation. This effect is not constant but changes over the time. The results of this study show that RS information could be successfully used as a primary data source for environmental planning. There is a strong need among policy makers for this type of information and a methodology such as developed here.

6. SUSTAINABLE DEVELOPMENT

Sustainable development is not a new concept. According to the U.S.A. Centre of Excellence for Sustainable Development, it is the latest expression of a long-standing ethic involving people's relationships with the environment, and the current generation's responsibilities to future generations. For a community to be truly sustainable, it must adopt a three-pronged approach that encompasses economic, environmental and cultural resources. Communities must consider these needs not only in the short term, but also in the long one. Several questions arise form this concept: In **Chapter 8** we described them as socio-economic and ecological subsystems. From this model we derive five main policies for optimising the relationships between both systems.

1. Ecosystem conservation management.
2. Economic optimisation management
3. Sustainable use of resources
4. Sustainable use of environmental services
5. Total system management (i.e. sustainable development).

A strategic question at the start of a regional sustainable development is: what is the size or complexity of the problems that sustainability poses to regional planning? In the volcanoes' region, problems are large and complex because the conservation of its natural capital is being endangered. At the current state of models, plans, policies and directives can be tailored to produce the desired effects, contributing directly to the decision making processes and policies design.

One of the great advantages of modelling is its theoretical ability to forecast the events resulting from the action of a system. For the researcher, a model is the means to experiment and verify his hypotheses by selective modification of the model and the input data. To the land manager, it gives the possibility to simulate the transformations an area will undergo when the conditions are changed. Another interesting point of simulations is the possibility of confronting their results with the real world, so that the experience gained can be reintroduced into the model.

In Chapter 8, we describe the strategic planning efforts applied to a recreational area of Iztaccíhuatl-Popocatépetl National Park and one of its outlying municipalities. The research was designed for the application of GIS tools and multi-criteria and multi-objective decision-making approaches. The results obtained in Chapter 8 met the demands posed by the problem. The scenarios rank highly the importance of forestry resources as self-generating media, producing economical and environmental benefits at the same time. On basis of this case study, a further design of management plans to reduce vegetation loss and fragmentation can be developed.

7.   NATURAL RESOURCE MANAGEMENT

The concept of Natural Resource Management is increasingly used as an analytical framework for applied ecology. The question is how does it broaden the scope of what managers and policy makers need to consider. The concept of Natural Resource Management is reflected in Figure 9.1. Human activities cause changes in the environment, which consequently cause changes in (eco-) systems. This can be among other activities such as burning, grazing, cutting, recreation, emission of pollutants, and extension of build-up areas. Such effects on ecosystems will almost certainly influence ecological processes and feedback mechanisms.

The functioning of any system can be described in terms of indicators or proxies, which are variables reflecting the structure and dynamics of the system. In this way, a particular effect of activities can be expressed in changed values of such indicators. System analysis and evaluation of effects through indicator values do help to characterize the condition of the system.

The desired state of the system can be defined in terms of targets, i.e. directive values of the indicator. In contrast to fundamental research into the processes of ecosystems, targets for indicators cannot be freely chosen. Targets represent the conviction of those who value the indicators. An important target for ecosystems is e.g. the species composition.

Taking into account the different (ecological and socio-economic) indicators, determining whether indicator values do accomplish the targets is part of a multi-criteria analysis. The result of such a multi-criteria analysis will result in a so-called pay-off table for each criterion given the conditions of the system and the activities/management measures taken.

Because of many uncertainties, scenarios are used to be able to analyse the sensitivity of the ecosystem under different conditions. Subjects of scenarios are for example autonomous changes in human pressure. Scenarios are either subject of autonomous activities and/or autonomous system processes. Thus the autonomous processes cannot be influenced by the decision making process.

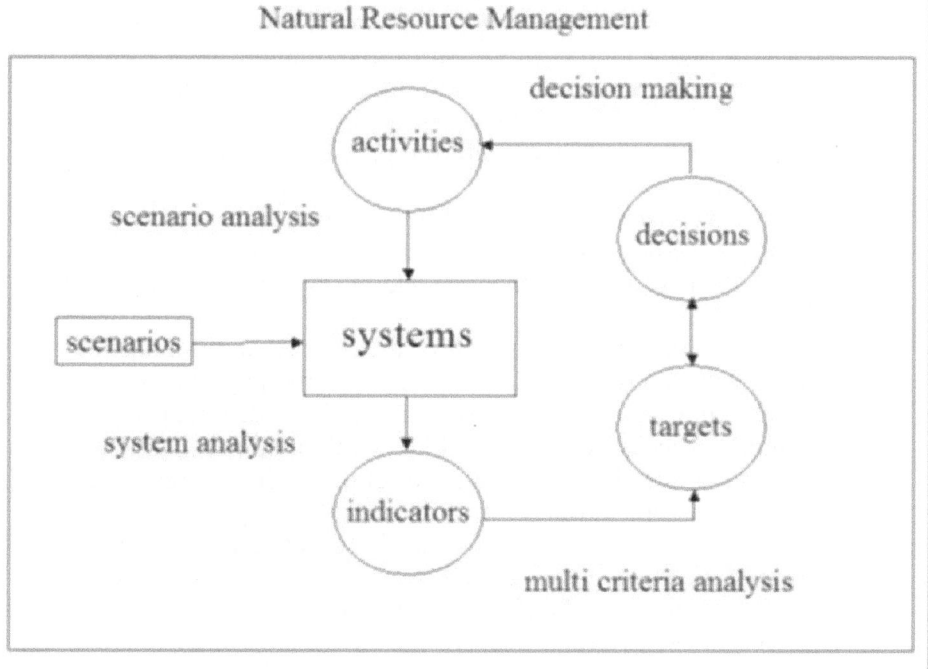

Figure 9.1. Concept of Natural Resource Management

The immediate effect of decision-making should be the implementation of management measures to improve the functioning of the system. This includes the ecological and socio-economic targets. It should be clear that there could be significant differences between sets of management measures that can accomplish the same combination of targets (e.g. Ridgley and Heil 1998).

We can consider Natural Resource Management as a conscious effort to manipulate factors to arrive at a desired system functioning. How should managers determine the best land use and land cover for an area? The procedure it should follow can be described as consisting of seven steps (*cf.* Ridgley and Heil 1998):

1. First, it needs to become clear about the targets one wants to achieve and to translate these requirements into operational criteria, proxies or indicators that one can use to judge if one proposed management measure is better than another.
2. Then, one should determine how a change on the system would affect those targets, e.g. by simulation modelling. Knowing the effects of these individual changes, the policy maker/manager will need to decide what mix of measures to promote.
3. Since different measures will yield different effects, the policy maker/manager (PM&M) must first decide on a desired balance of targets achievement. This "preferred payoff mix" now becomes the managers' overall goal for the area.
4. The PM&M designs a range of different management plans, each corresponding to a different set of targets that represents a different development scenario for the area and the region within which it falls.
5. The PM&M now search for a set of measures that minimizes conflict between the plan yielding its preferred payoff mix and those corresponding to the different scenarios. Since different sets of management measures may yield identical or similar payoffs, the manager may be able to formulate a set that will accomplish their objectives under a variety of futures.
6. The PM&M now must determine the best location and spatial pattern of the changes prescribed by each minimum-conflict set. This is the task of *landscape planning*. In effecting it, the PM&M's will consider structural aspects of the landscape as they relate to the ecological, management, and other criteria. GIS is the principal tool employed for this activity. The outcome is a set of alternative minimum-conflict landscape plans.

7. As the final step, the PM&M makes a decision on a course of action. Two broad strategies suggest themselves. The PM&M can evaluate the alternative plans and select one to adopt and promote. Alternatively, the PM&M could identify the land-use changes common to all or many of the plans and take actions to implement those changes as soon as possible. Those changes constitute a "commitment set". As the future development pattern around the protected area becomes clearer, the manager could identify further changes, perhaps repeating the above procedure.

8.   CONCLUSIONS

In this Chapter, we have highlighted our results in the perspective on how we should think about ecology and man. In any one place, the species present in the ecosystem will be there as a result of the complex interaction of a number of abiotic and biotic factors, and the actions of man over space and time. A restoration project following a disturbance of some kind will be able to return the same set of species only by replicating the exact combination of process responses that created the pre-existing ecosystem.

In a way, environmental factors simply represent ecological variables and processes that are likely to be important in any given situation. However, by formalising our understanding of the likely importance of a range of these variables, we are able to influence those that are likely to be important determinants of restoration success in any given place (*cf.* Hobbs and Norton, 2003).

In addition to ecological factors, also a set of socio-economic factors will be important, i.e.:

a. Wants: i.e., the goals and aspirations of local communities, land owners and other stakeholders -e.g. alternative goals relating to biodiversity conservation, watershed protection, soil conservation or maximising production

b. Costs: i.e., the costs relating to achieving stated goals in relation to production, planting, and maintenance.

An important indicator for conservation and restoration targets is the species composition. If human impact can be decreased then succession can be arrested or set back to an earlier stage of development. Restoration principles are mainly based on (Lockwood and Samuels 2003):

Passive restoration: The first step to restoration is to arrest degradation activities that prevent recovery. Understanding the system-specific steps in natural succession can inform management decisions. Monitoring and observing the natural recovery process will show whether passive restoration might work. Beyond saving time and money, this approach can inform management decisions if active restoration methods are required (Kauffman *et al.* 1997).

Interseeding: Interseeding involves sowing selected plant species in with existing organic soil material. This approach essentially allows nature to initiate succession. This simple measure improves the chances of incorporating conservative species in the restored vegetation. However, the development may not be linear, so that caution must be applied in determining where one stands among the different community assembly pathways (Hobbs and Norton 1996).

Re-introduction of keystone species: Re-introductions of species must be well planned in accordance with successional processes. The timing of introduction of a keystone species is critical. It is also something we can deduce from an examination of a reference system. Identifying these species or knowing when they will do more harm than good is not as easy as one could think of. Sources of information are likely found by a close examination of the surrounding landscape.

However, it is obvious that ecosystems are complex dynamic systems with the tendency to be present in different states. Transitions between these states may be rapid, and there may be thresholds, which, once crossed, makes movement back to previous states difficult. The probability for different ecosystem states is an important input for natural resource management, since it means that we must be clear about the targets we set for projects and how stringently we are going to demand the return to a particular state, in relation to our ability to manage the system. It is here where ecology and man must be integrated into one system.

REFERENCES

Braun-Blanquet, J.J. (1964). Pflanzensoziologie, Grundzüge der Vegetationskunde. 3rd Edition. Springer, Vienna, New York, 865 pp.

Boo, E. (1993) Ecotourism planning for protected areas.15-31p. In: K. Lindemberg and D. E. Hawkings (Eds.). Ecotourism: a guide for planners & managers. The Ecotourism Society, North Bennington, Vermont.

Chávez, J.M. and Trigo, N. (eds.) (1996). Programa de Manejo para el Parque Nacional Iztaccíhuatl-Popocatépetl. Colección Ecología y Planeación, Universidad Autónoma Metropolitana, Xochimilco, México. 273 pp. 12 mapas.

Hobbs, R. J. and Norton, D. A.1996. Towards a conceptual framework for restoration ecology. Restoration Ecology 4: 93-110.

Hobbs, R.J. and Norton, D. A.2003. Ecological filters, thresholds and gradients in resistance to ecosystem reassembly. In: V. Temperton, R. Hobbs, T. Nuttle and S. Halle (eds.) Assembly Rules and Restoration Ecology – Bridging the Gap Between Theory and Practice. Island Press.

Kauffman, J.B., Beschta, R.L. Otting, N.and Lytjen, D. 1997. An ecological   perspective of riparian and stream restoration in the western United States.   Fisheries 22(5): 12-24.

Lockwood J.L. and Samuels C.L. (2003). Incorporating assembly theory into the practice of restoration. In: V. Temperton, R. Hobbs, T. Nuttle and S. Halle (eds.) Assembly Rules and Restoration Ecology – Bridging the Gap Between Theory and Practice" . Island Press.

Ridgley, M. and Heil, G.W. (1998) Multicriterion planning of protected-area buffer zones: an application to Mexico's Izta-Popo National Parl. In: Beinat E and Nijkamp P (eds) Multicriteria Analysis for Land-Use Management, pp 293–311. Kluwer Academic Publishers, Dordrecht, The Netherlands

Index